3rd International Summer School and Workshop on Complex and Magnetic Soft Matter Systems: Physico-Mechanical Properties and Structure (CMSMS'17)

Journal of Physics: Conference Series Volume 994

Dubna, Russia
28 - 30 June 2017

Editors:

Yuriy Raikher
Maria Balasoiu
Andrey Rogachev
Madalin Bunoiu

ISBN: 978-1-5108-6071-1
ISSN: 1742-6588

Printed from e-media with permission by:

Curran Associates, Inc.
57 Morehouse Lane
Red Hook, NY 12571

Some format issues inherent in the e-media version may also appear in this print version.

Printed by Curran Associates, Inc. (2018)

For permission requests, please contact the Institute of Physics
at the address below.

Institute of Physics
Dirac House, Temple Back
Bristol BS1 6BE UK

Phone: 44 1 17 929 7481
Fax: 44 1 17 920 0979

techtracking@iop.org

Additional copies of this publication are available from:

Curran Associates, Inc.
57 Morehouse Lane
Red Hook, NY 12571 USA
Phone: 845-758-0400
Fax: 845-758-2633
Email: curran@proceedings.com
Web: www.proceedings.com

Table of contents

Volume 994

3rd International Summer School and Workshop "Complex and Magnetic Soft Matter Systems: Physico-Mechanical Properties and Structure"

28–30 June 2017, Dubna, Moscow Region, Russian Federation

Accepted papers received: 19 March 2018
Published online: 28 March 2018

Preface

Papers

3rd International School and Workshop on Complex and Magnetic Soft Matter Systems: Structure and Physico - Mechanical Properties

Organized by

Joint Institute for Nuclear Research, Dubna
West University of Timisoara
Institute of Continuous Media Mechanics of Russian Academy of Sciences, Perm
Belarusian National Technical University, UNESCO Chair
Horia Hulubei National Institute of Physics and Nuclear Engineering, Bucharest
Romanian Society of Physics

Supported by

The RO-JINR Grant №218/10.04.2017 item 14, 15 and
RO-JINR Projects № 219/10.04.2017 item 62, 63, 65

Guest Editors

Yuriy Raikher, Maria Balasoiu, Andrey Rogachev, Madalin Bunoiu

International Advisory Committee

Ioan Bica (WUT, Timisoara) *(Chair)*; Yuriy Raikher (IMCC, Perm) *(Co-Chair)*;
Victor Bashtovoi (BNTU, MINSK) *(Co-Chair)*; Peter Apel (JINR, Dubna);
Alexander Belushkin (JINR, Dubna); Leonid Bulavin (NTSU, Kiev);
Emil Burzo (CESS, Cluj-Napoca); Andrejs Cebers (Latvian University, Riga);
Valentin Gordeliy (IBS, Grenoble; JINR, Dubna); Michal Hnatic (JINR, Dubna);
Christian Holm (ICP, Stuttgart); Alexei Khokhlov (MSU, Moscow);
Peter Kopchansky (IEP, Kosice); Alexei Kuzmin (ISSP, Riga);
Norbert Kucerka (JINR, Dubna); Nicoleta Lupu (IFT, Iasi);
Alexander Ozerin (ISPM, Moscow); Regine Perzynski (Universite P&M Curie, Paris);
Sergey Tiutiunikov (JINR, Dubna); Irina Tchugueva (RAS, Moscow);
Jose Teixeira (LLB, Saclay); Jan Wasicki (AMU, Poznan)

Programme Committee

Yuriy Raikher (ICMM, Perm) *(Chair)*; Maria Balasoiu (JINR, Dubna/IFIN-HH,Bucharest);
Madalin Bunoiu (WUT, Timisoara); Regine Perzynski (U P&M C, Paris);
Sergey Tiutiunikov (JINR, Dubna)

Organizing Committee

Maria Balasoiu *(Chair)*; Andrey Rogachev *(Co-Chair)*;
Tatiana Murugova *(Scientific secretary)*;
Oleksandr Ivankov; Dorota Chudoba; Yulia Gorshkova;
Alexander Kuklin; Dmitro Soloviov; Maxim Rulev;
Sergey Kozenkov; Ekaterina Efimova; Peter Melenev;
Anastasia Biller; Alexander Rhyzhkov; Svetlana Chubakova;
Natalia Dokalenko *(Conference secretary)*

IOP Conf. Series: Journal of Physics: Conf. Series **994** (2018) 011001 doi:10.1088/1742-6596/994/1/011001

CMSMS'17 - 3rd International School and Workshop on Complex and Magnetic Soft Matter Systems: Structure and Physico -Mechanical Properties

Yuriy Raikher[1,2*], **Maria Balasoiu**[3,4*], **Andrey Rogachev**[3,5*], **Madalin Bunoiu**[6*]

[1]Institute of Continuous Media Mechanics, Perm Federal Research Center, Ural Branch of RAS, Perm, Russia
[2]Institute of Natural Sciences and Mathematics, Ural Federal University, Ekaterinburg, Russia
[3]Joint Institute for Nuclear Research, Dubna, Russia
[4]Horia Hulubei National Institute of Physics and Engineering, Bucharest, Romania
[5]Moscow Institute of Physics and Technology (MIPT), Dolgoprudny, Russia
[6]West University of Timisoara, Timisoara, Romania

*Guest Editors

Preface

The CMSMS'17 was organized by Joint Institute for Nuclear Research (Dubna), West University of Timisoara, Institute of Continuous Media Mechanics of Russian Academy of Sciences (Perm), UNESCO Chair of the Belarusian National Technical University, Institute of Physics and Nuclear Engineering "Horia Hulubei" (Bucharest) and Romanian Society of Physics.

The topics of lectures, oral and poster presentations covered the theory, simulations, and experimental research in physical, mechanical, structural and chemical aspects of the materials science as well as biological aspects of soft complex matter with special emphasis on soft magnetic matter (magnetic elastomers, ferrogels, ferroliquid crystals, associations of nanoparticles with biomolecules and cells, etc.).

The first International Summer School and Workshop "Complex and Magnetic Soft Matter Systems: Physico-Mechanical Properties and Structure" (CMSMS'12) took place from 3 to 7 September 2012 in Alushta. It has brought together 32 specialists (top-level experts and young researchers) from Russia, Romania, Moldova, Belarus, Ukraine, France and Germany working on interdisciplinary problems in soft matter. Their contributions totaled 37 works.

The second event (CMSMS'14) was organized in Dubna from 29 September to 3 October 2014. It has brought 65 participants from Russia, Romania, Poland, Slovakia, Germany, and France. More than 10 invited speakers from Russian and foreign institutions provided a brief overview of several important research issues in complex and magnetic soft matter fields. Young scientists demonstrated a high level

IOP Conf. Series: Journal of Physics: Conf. Series **994** (2018) 011001 doi:10.1088/1742-6596/994/1/011001

at both the poster session and the oral presentation. The programme included an excursion to the Frank Laboratory of Neutron Physics, where the participants visited the high-pulsed reactor IBR-2 and the intense resonance neutron source IREN.

The main objective of CMSMS'17, organized in Dubna from 28 to 30 June 2017, was to exchange the latest research findings in the scope of the meeting. Around 60 scientists from Azerbaijan, Belarus, Germany, Poland, Romania, Russia, and Slovakia have participated. A number of 14 invited lectures delivered concise overviews of actual research issues in the school topics. Young scientists were encouraged to present their results in oral talks. The poster sessions extended the frame of oral sessions providing opportunities for the scientific exchange in personal discussions to be continued. The presentations amounted to 35 contributions.

IOP Conf. Series: Journal of Physics: Conf. Series **994** (2018) 011001 doi:10.1088/1742-6596/994/1/011001

Conference Photos

3rd International School and Workshop on Complex and Magnetic Soft Matter Systems: Structure and Physico -Mechanical Properties, 28-30 June 2017, Dubna

IOP Conf. Series: Journal of Physics: Conf. Series **994** (2018) 011001 doi:10.1088/1742-6596/994/1/011001

During the lectures

IOP Conf. Series: Journal of Physics: Conf. Series **994** (2018) 011001 doi:10.1088/1742-6596/994/1/011001

During the lectures

IOP Conf. Series: Journal of Physics: Conf. Series **994** (2018) 011001 doi:10.1088/1742-6596/994/1/011001

During the poster session

IOP Conf. Series: Journal of Physics: Conf. Series **994** (2018) 011001 doi:10.1088/1742-6596/994/1/011001

In between sessions

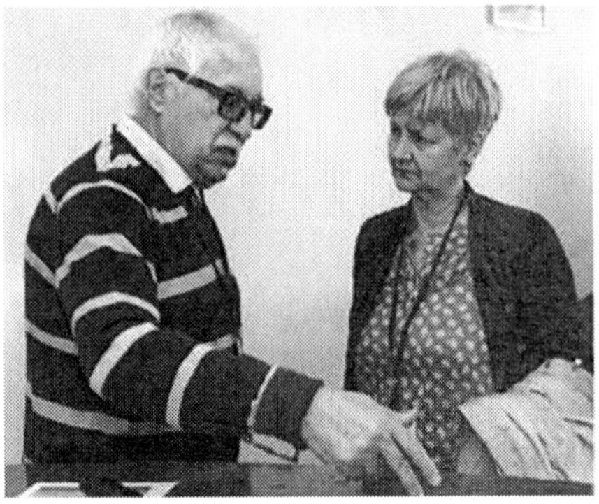

Acknowledgements

The Organizers of the 3rd International School and Workshop on Complex and Magnetic Soft Matter Systems: Structure and Physico -Mechanical Properties acknowledge the financial support from the Grants of the Government Plenipotentiary Representative of Romania and scientific projects of the Cooperation Programmes JINR-Romania. Special gratitude to the JINR and FLNP administration and staff is expressed.

Peer review statement

All papers published in this volume of *Journal of Physics: Conference Series* have been peer reviewed through processes administered by the proceedings Editors. Reviews were conducted by expert referees to the professional and scientific standards expected of a proceedings journal published by IOP Publishing.

IOP Conf. Series: Journal of Physics: Conf. Series **994** (2018) 012001 doi:10.1088/1742-6596/994/1/012001

Two-particle element of a magnetorheological elastomer under a cyclic magnetic field

A M Biller, O V Stolbov, Yu L Raikher

Institute of Continuous Media Mechanics – the Division of Perm Federal Research Center, Russian Academy of Sciences, Ural Branch, Perm, 614013, Russia

E-mail: kam@icmm.ru

Abstract. The results of modelling the behaviour of a pair of magnetisable spherical particles embedded in a viscoelastic elastomer with Kelvin rheology are presented for the case, where an ac magnetic field is exerted along the intercenter line of the particles. This system is considered as a small-scale (mesoscopic) structure element of a magnetorheological elastomer. The system in question is known to be able for hysteretic behaviour under the quasistatic cycle of the field. The particles initially positioned well apart, at some finite field strength, fall onto one another (cluster) and dwell in this state until the field decreases well below the value at which the cluster has been formed. Under dynamic cycling of the field, the viscous friction interferes the particle displacement process and impedes occurrence of the magnetodeformational hysteresis of the element. Starting from small-amplitude overdamped oscillations and enhancing the role of magnetic forces over viscous ones, we show how the anharmonicity of the system grows while it approaches the transition threshold, above which the dynamics of the element includes the cluster state. The dynamic hysteresis of magnetisation that accompanies the mechanical oscillations of the particles is presented as well.

1. Introduction

Among the many newly developed materials, the so-called "smart" ones attract special interest because of their properties which can be controlled and adjusted to variable external conditions. Such materials include magnetorheological elastomers (MRE), the composites consisting of an elastomer matrix filled with micron-sized magnetisable particles. A distinctive feature of these materials is their magnetosensitivity: when an external magnetic field is applied, the particles inside the composite magnetise and interact with each other changing their spatial distribution. This causes modification of a number of the MRE macroscopic properties: elasticity and viscosity, magnetic and electric permeability, conductivity, etc. Usage of MRE as essential working components in adaptive dampers, actuators, valves, medical appliances, etc. is an actively developing applicational trend. In such devices, the main exploited effect is the change of rheological characteristics and shape of MRE elements in response to the applied magnetic field or mechanical load in the real-time regime. Therefore, the dependence of the dynamic characteristics of MREs on the applied field and load is an important issue; this is demonstrated by a rather large number of published experimental works on this subject [1, 2, 3, 4, 5, 6].

2. Two-particle element

2.1. *Quasistatic situation*

To understand the mechanisms of inner interactions in MREs, which govern rearrangement of their internal structure, one should study them at the mesoscopic level, where the matrix and particles are considered separately and each one is treated as a continuous medium with its own set of properties to which scope the hypotheses on the interaction between the phases are added. In line with this viewpoint, the present paper considers a set of two magnetic particles embedded in an elastic matrix as a basic element of an MRE capable of illustrating the magnetomechanical response of the material. This model element consists of a cylindrical sample whose outer dimensions are chosen under the requirement that a pair of spherical particles occupies about 30% of its geometric volume. The particles have identical radii a, and are positioned at the cylinder axis in such a way that their intercenter vector $\boldsymbol{l}$ lies along this axis. The latter is taken as the Oz axis of the cylindrical coordinate frame whose origin is placed at the centre of the sample.

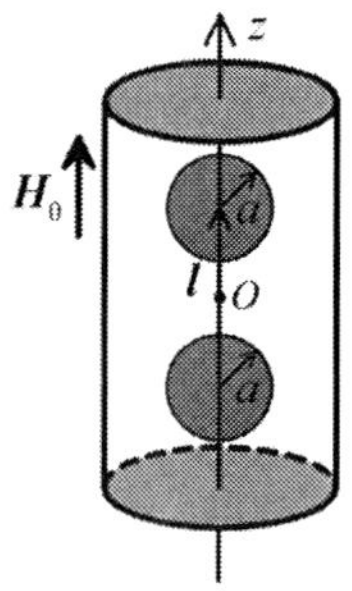

Figure 1. Model two-particle element of an MRE.

As well known, the most strong shape and magnetic effects are displayed by MREs based on soft polymer matrices with Young moduli $E \sim 5-30$ kPa, see [4], for example. The internal viscosity of those compounds estimated from their relaxation times which are of the order of a few seconds [7]. This time scale is inherent to MREs filled with microparticles (2–5 μm) of carbonyl iron that has the initial magnetic susceptibility is of the order of 10^4 and saturation magnetisation of 1500 kA/m. In view of what follows, we note that under not too strong fields and strains, the nonlinearity of both the particle magnetisation and the matrix elasticity might be neglected, at least in the first approximation.

When the element sketched in Fig. 1 is subjected to an external uniform magnetic field $\boldsymbol{H}_0 = (0, 0, H_0)$, the magnetic polarisation of the particles induces their mutual attraction striving to move them closer. However, as soon as the particles yield to this tendency, the deformed matrix generates restoring elastic forces, so that a new equilibrium state of the pair is defined by the balance of those opposing effects.

The magnetomechanical response of the above-described two-particle element under quasistatic conditions was studied in detail in Refs. [8, 9]. In that modelling, nonuniform deformation of a nonlinearly elastic elastomer matrix and nonuniform and nonlinear magnetisation of the particles were taken into account as well. It turned out that within a certain domain of material parameters of the particles and matrix, the system displays bistable behaviour. Namely, when the MRE element is subjected to a quasistatic cycle of the magnetic field, the magnetomechanical hysteresis occurs: the path that the particles pass under the increase of the field does not coincide with that undergone during the field decrease. The main parameter governing the effect is $\beta = \mu_0 M_s^2/c_1$, where μ_0 is the magnetic constant, M_s saturation magnetisation of the ferromagnetic material, and c_1 elastic constant of the Mooney-Rivlin model assumed for the polymer.

Further on, to characterise the configuration of the two particle element, we use nondimensional units, scaling all the distances with the particle radius. Therefore, the current intercenter distance of the pair and the relative change of the interparticle gap (the closest distance between the surfaces of the particles) are defined as

$$q = l/a, \quad \text{and} \quad \epsilon = (q-2)/(q_0-2), \tag{1}$$

respectively; here q_0 is the initial interparticle distance and $q = 2$ its limiting value attained when the particles come into tight contact. As seen, the gap parameter ϵ varies from unity in the initial state of the element to zero when the particles collapse in a pair cluster.

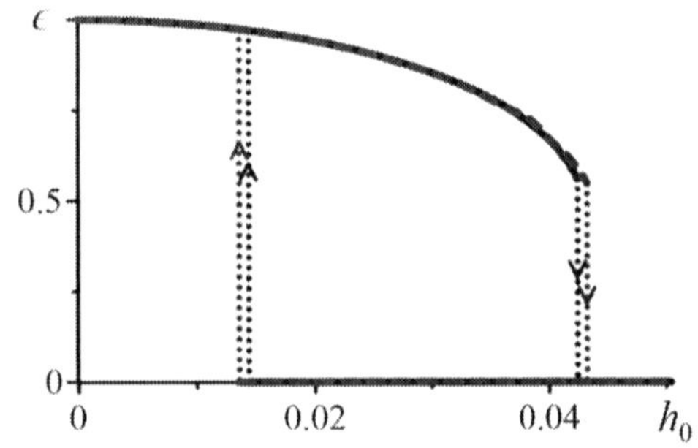

Figure 2. Quasistatic magnetomechanical hysteresis loops for a pair of particles with intercenter distance $q_0 = 3$; solid lines correspond to the case of particles made of a linearly magnetisable ferromagnet, dashed lines are for the case of a ferromagnet that magnetises according to the Frölich-Kennelly law [10] with the parameters given in the text. Note virtual coincidence of the curves of both types, the difference is visible only in the vertical lines denoting the jumps to and from the cluster state

In Fig. 2 an example of quasistatic bistable behaviour is presented for the element with initial intercenter distance $q_0 = l_0/a = 3$ and magnetomechanical parameter $\beta = 2500$, that corresponds to a MRE with $M_s = 1500\,\text{kA/m}$ and Young modulus of the matrix 8 kPa. Here and below we express the external field strength in nondimensional units as $h_0 = H_0/M_s$. The adopted set of material parameters implies that the polymer is rather soft so that the collapsing field is low: $h_0 \approx 0.04$ that corresponds to $H_0 \sim 60\,\text{kA/m}$ (or 750 Oe) that is very far from the saturation level. That is why the magnetomechanical hysteresis loops of the linearly and nonlinearly magnetising particles do not differ much. Further on, being after qualitatively understanding of the dynamic processes, we restrict the consideration by linear magnetisation approximation using it for evaluating the interparticle magnetic interaction via the *linear multipole solution* model. This treatment is not at all unrealistic since iron carbonyl microparticles magnetise virtually linearly in the fields up to few kOe. Although a linearly magnetising substance does not have saturation, in below we keep the adopted value of M_s as the unit of the field scale; this should be recalled when transforming the results back to the dimensional form.

2.2. *Time-dependent formulation*

In the case of a time-dependent field, the motion of the particles, besides the magnetic and elastic forces, which are potential, is affected by viscous friction which has two evident sources. One stems from the interaction of eddy currents induced inside the metal grains by the applied field.

IOP Conf. Series: Journal of Physics: Conf. Series **994** (2018) 012001 doi:10.1088/1742-6596/994/1/012001

However, in the low-frequency (up to few kHz) range that we aim at, this effect is negligible. The dissipation mechanism that is relevant in the considered situation results from the particle movement relative to the matrix. Provided the motion is not too fast, the arising viscous force acting on a particle is Stokes-like, i.e., proportional to its velocity.

We begin with the case of small particle displacements, i.e., small matrix strains. In that case, one may assume that the particles move only along the upper branch of the magnetomechanical hysteresis loop, see Fig. 2. In linear-strain approximation, the stress $\boldsymbol{\sigma}$ arising in the matrix can be represented as a sum of elastic and viscous parts as

$$\boldsymbol{\sigma} = \boldsymbol{\sigma}_{el} + \boldsymbol{\sigma}_{visc} = \lambda I_1(\boldsymbol{\varepsilon})\boldsymbol{I} + 2G\boldsymbol{\varepsilon} + \eta\dot{\boldsymbol{\varepsilon}}. \tag{2}$$

Here $\boldsymbol{\varepsilon}$ is the strain tensor, λ is the Lamé's first parameter, shear modulus G and viscosity η of the material. The adopted representation corresponds to the Kelvin-Voigt model of a viscoelastic medium, where the elastic and viscous elements are connected in parallel.

Under linear magnetisation law, the interparticle magnetic force may be described with the aid of analytical expressions derived in Ref. [8]. In the here adopted units the result of [8] for the magnetic attraction force per particle writes as

$$F_m = \mu_0 H_0^2 a^2 \tilde{F}_m(q), \tag{3}$$

where nondimensional function $\tilde{F}_m$ in a complicated way depends on the interparticle distance q but is independent of the applied field strength. Introducing volume density $\boldsymbol{f}_m$ of this force in such a way that it is nonzero only inside the particles and is always directed along the local magnetic field, one arrives at the time-dependent force balance equation

$$\nabla \cdot \boldsymbol{\sigma} + \boldsymbol{f}_m = 0. \tag{4}$$

Here the inertia terms are omitted as being negligible, see Section 6 for estimates.

Finally, we introduce the dimensionless time scale $\tilde{t} = t/\tau_R$ basing it on the stress relaxation time $\tau_R = \eta/G$ of the Kelvin-Voigt model. Transformation of the time derivatives $\dot{\tilde{\boldsymbol{\varepsilon}}} = \tau\dot{\boldsymbol{\varepsilon}}$ yields the stess tensor in the form

$$\boldsymbol{\sigma} = G\tilde{\boldsymbol{\sigma}} = G\left[\frac{\lambda}{G} I_1(\boldsymbol{\varepsilon})\boldsymbol{I} + 2\boldsymbol{\varepsilon} + \dot{\tilde{\boldsymbol{\varepsilon}}}\right], \tag{5}$$

so that the equation of motion (force balance) writes as

$$\tilde{\nabla} \cdot \tilde{\boldsymbol{\sigma}} + \kappa h_0^2 \tilde{\boldsymbol{f}}_m = 0, \tag{6}$$

where $\tilde{\nabla} = a\nabla$ and the coefficient alongside the magnetic force is

$$\kappa = \mu_0 M_s^2 a^3/G. \tag{7}$$

The magnetoelastic problem under study is completed by adding to it the kinematic relations

$$\boldsymbol{\varepsilon} = \frac{1}{2}\left[(\nabla \boldsymbol{u})^T + (\nabla \boldsymbol{u})\right] \tag{8}$$

(where the superscript denotes matrix transposition) and stress-free conditions for all boundaries.

3. Oscillations of the interparticle distance

Let an ac uniform external magnetic field

$$h_0 = h_A \sin(\omega t) = h_A \sin(\tilde{\omega}\tilde{t}), \qquad \tilde{\omega} = \omega\tau_R \tag{9}$$

of amplitude h_A be applied along the axis of the MRE element; here ω is dimensional frequency. For the steady field oscillations (9), the magnetomechanical problem (5)–(8) was solved numerically in axisymmetric formulation by the finite element method with the aid of library fenics for the python language.

As the magnetic particles are fully polarisable and do not possess any residual magnetisation, the magnetic force between them does not depend on the direction of the applied field and, thus, would oscillate with doubled frequency. Accordingly, the same would be the time dependence of the forced changes of the interparticle gap ϵ.

3.1. Low-field amplitudes

Let us compare solution of the dynamic problem with the results of quasistatics. For that, we consider the field-induced oscillatory solutions of the set (5)–(8) and vary the frequency. The corresponding plots of the interparticle distance $\epsilon(h_0)$ are presented in Fig. 3.

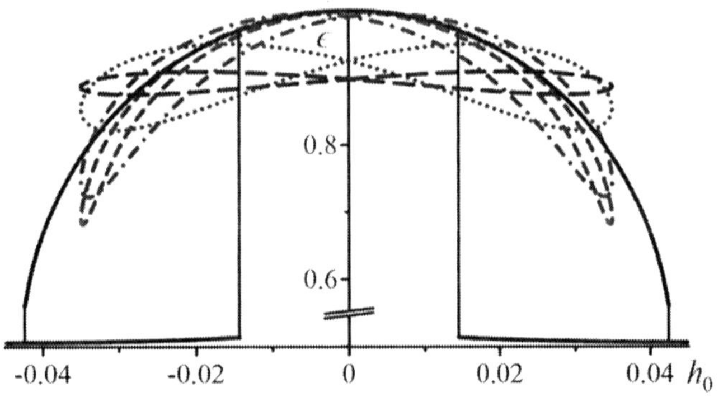

Figure 3. Interparticle distance as a function of the applied field strength in the quasistatic case (solid line), and in steady oscillatory regimes induced by ac field of amplitude $h_A = 0.035$ and frequencies $\tilde{\omega} = 0.03$ (dashed), 0.15 (dash-dotted), 0.75 (dotted), and 3.75 (long dashed)

For comparison, in Fig. 3 the quasistatic loop same as in Fig. 2 is presented having been expanded with allowance for the fact that the field that may assume the values of both signs, as it happens for the ac field. However, the dynamic cycles in Fig. 3 are calculated for the field whose amplitude does not exceed the quasistatic threshold value, and, thus, no jumps of the interparticle distance occur. This means that the shown loops are entirely due to the dynamic hysteresis – the viscosity-induced lag between the field (the cause) and the displacement (the response) – because under this condition the system performs cycles near the upper branch of the quasistatic magnetodeformational curve and never fall in the cluster state. Indeed, at the lowest frequency, the loop is narrow (almost adiabatic) and closely follows the quasistatic one, see the dashed line. With the increase of frequency, the loop becomes wider which effect is maximal when the dispersion condition $\tilde{\omega} \sim 1$ is attained, see the dotted line in Fig. 3. Upon further growth of frequency, the response goes down as the exciting field oscillates too fast to entrain the particles to motion; accordingly, the loops shrink and become more “horizontal”. They also become more “shallow”: the minimal values, which they attain, progressively increase, less and less differing from unity that means that the span of the particle displacements diminishes.

For a given frequency ω, the greater viscosity, i.e., τ_R, the lower parameter $\tilde{\omega}$, the smaller number of cycles the system has to undergo between the initial state and attaining a stable limiting cycle. This is illustrated by Fig. 4.

Note that the specific "baglike" shape of the loop at the lowest frequency. For that curve, the displacement swing is maximal, and for that regime the nonlinear dependence of the magnetic force, see (3), manifests itself. Indeed, the magnetic attraction force even in the simplest (dipole) approximation grows as $1/q^3$ when the particles approach. In the multipole solution, which takes into account mutual polarisation of the particles, this dependence enhances substantially (the power exponent increases) at closer distances, see [8]. That is why the same increment of h_0^2 – we remind that the external driving force is proportional to the square of the field – induces larger displacements. Due to that, the right part of the loop in Fig. 4a is wider that its left part. As Figs. 4b and c show, this effect fades out fast as soon as the particles displacement amplitude becomes smaller.

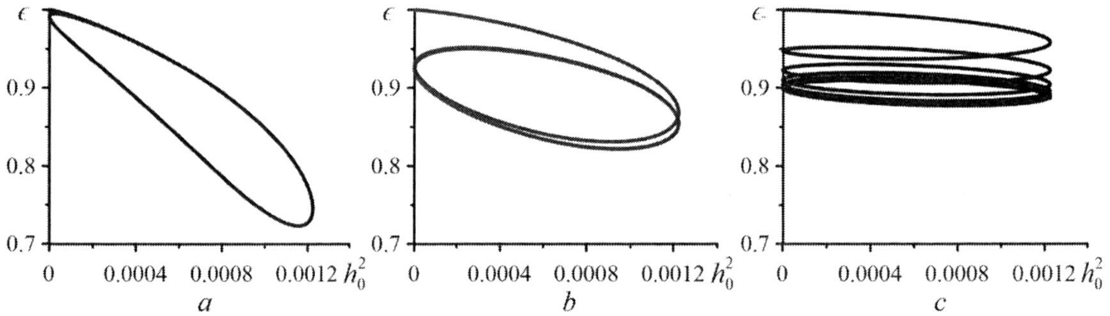

Figure 4. Establishing the steady dynamic magnetodeformational hysteresis cycle of the pair for the field amplitude $h_A = 0.035$ under frequencies $\tilde{\omega} = 0.15$ (a), 0.75(b) and 3.75(b)

3.2. Oscillations in pretransitional region

Evidently, the most interesting case of the problem under study are the oscillations of the particle pair, when the ac field is strong enough to provoke the particle collapse. During some part of the field period, where the field strength is the highest, despite the viscous impediment, the particles would dwell in the cluster state falling on each other at the beginning of this time interval and falling apart at its end. The chances to obtain this regime analytically, due to its strong nonlinearity, are very poor, and numerical modelling seems the only possible way to describe it. Unfortunately, the numerical tools used here are inept for an adequate description of this field-frequency domain: the employed calculation procedure becomes unstable there. Nevertheless, even with the present technique, we are able to get some notion of the occurring transition.

The results of those tests are shown in Fig. 5. Unlike Fig. 4, here the amplitude of the applied field exceeds the quasistatic threshold h_* of the particle collapse, cf. Fig. 2. In this situation, the transitions of a separated pair into cluster and back become possible. However, the fact whether or not such a transition would really occur depends on the frequency of the field. Indeed, in the presence of viscous friction, the transition to the cluster state, i.e., a substantial change of the interparticle distance, needs a finite time τ_*. If at a given frequency in each half-period of the ac field, the condition $h_0(t) > h_*$ holds for a time interval longer than τ_*, then the particle clusterisation should take place. If, to the contrary, this interval is shorter that τ_*, clusterisation is impossible. Although the time parameter τ_* is not calculable analytically, the numerical plots of Fig. 5 for the certain amplitude imply that $0.5 < \tau_* < 0.6$. This follows from the difference of the cycle shapes; the one at higher frequency is closed and located at $\epsilon \geq 0.5$, while the one at lower frequency – within the framework of present work – falls down to $\epsilon \ll 1$ and never returns.

In our viewpoint, the above-presented findings imply that the forced oscillations of the pair where cluster formation takes place are an interesting issue, whose solution would be useful for shading light on the internal dynamics of MREs. Because of that, this work deserves further investigation, both analytical and numerical.

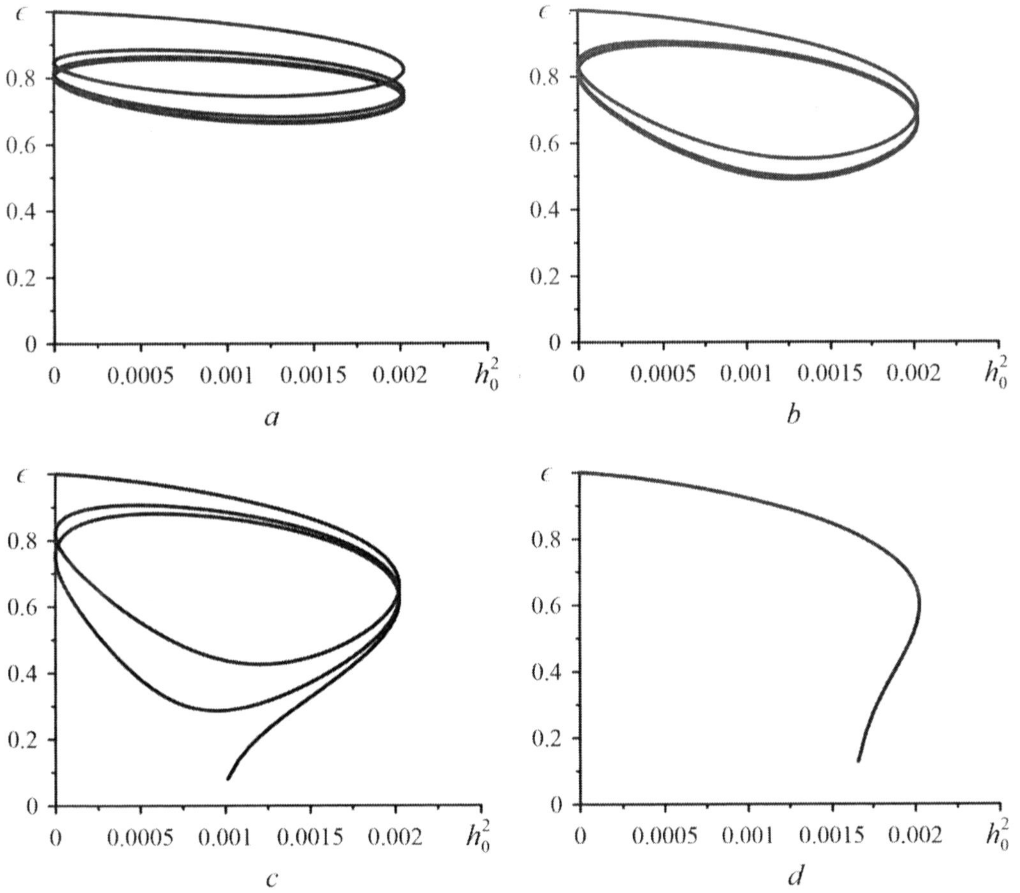

Figure 5. Steady magnetization curves of the model element under ac field of amplitude $h_A = 0.045$ and frequencies $\tilde{\omega} = 1.2$ (a), 0.6 (b), 0.5 (c), and 0.4 (d)

4. Magnetisation oscillations

As the MRE element under study is made of the magnetically polarisable particles of a finite size, the changes of interparticle distance modulate their mutual magnetisation. In result, the magnetic moment of the element undergoes cyclic oscillations. This effect could be evaluated from the carried out calculations and enables one to obtain the dynamic hysteresis loops of the model system under study.

In Fig. 6 plots of the magnetisation of the particle are presented for a constant subcritical ($H_0 < h_*$) amplitude of the ac field and different frequencies. Note that, due to the fact that in an MRE containing only magnetically soft particles, the shape of the dynamic magnetisation curves $m(h_0)$ look different in comparison with customary ones. The loops display two symmetrical

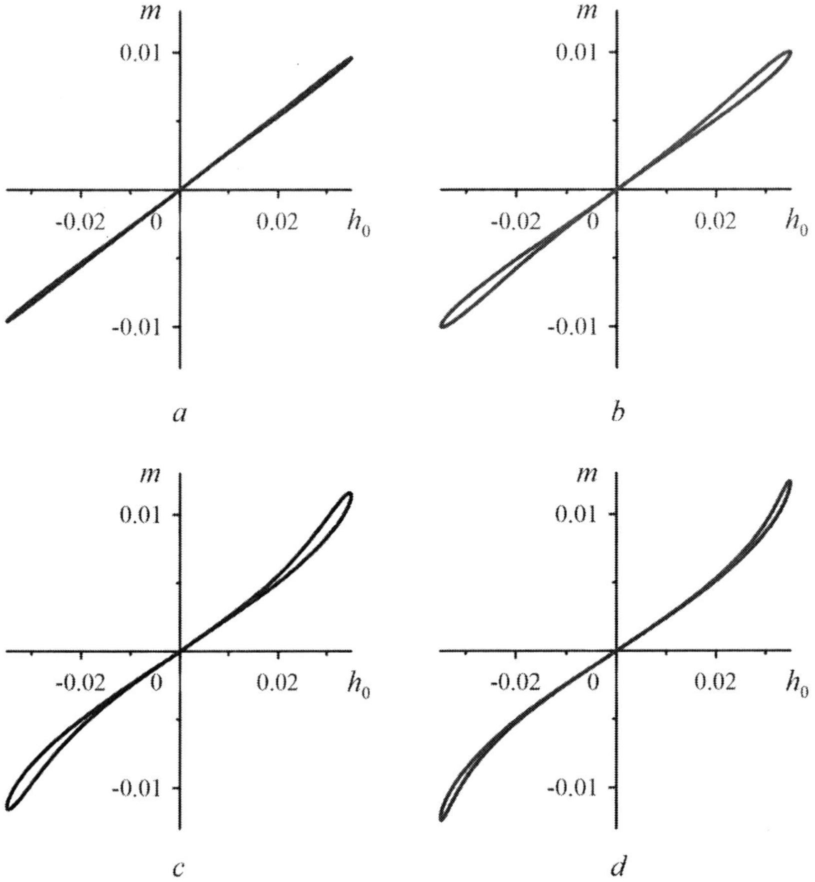

Figure 6. Steady magnetization curves of the model element under ac field of amplitude $h_A = 0.035$ and frequencies $\tilde{\omega} = 3.75$ (a), 0.75 (b), 0.15 (c), and 0.03 (d)

"inflation" regions at the ends, which are the result of the viscous friction. As it should be, the area of the loops has the maximum at $\tilde{\omega} = \omega\tau_R \sim 1$, where the energy dissipation due to the particle mechanical relaxation is the highest.

5. Conclusion

Modelling of the behaviour of a pair of magnetisable spherical particles embedded in a viscoelastic elastomer with Kelvin rheology is presented for the case, where an ac magnetic field is exerted along the intercenter line of the particles. This system might serve as a small-scale (mesoscopic) structure element of a magnetorheological polymer. A fundamental fact is that the system in question displays hysteretic behaviour under the quasistatic cycle of the field provided the latter is strong enough. The magnetic particles, initially positioned well apart, at some finite field strength fall onto one another (cluster) and dwell in this state until the field decreases well below the value at which the cluster has been formed. Under dynamic cycling of the field, the viscous friction interferes the particle displacement process and, provided the frequency of the field is high enough, impedes occurrence of the magnetodeformational hysteresis of the element. Starting from small-amplitude overdamped oscillations and enhancing the role of magnetic forces over viscous ones, we show how the anharmonicity of the system grows while it approaches the transition threshold, above which the dynamics of the element includes the cluster state. The dynamic hysteresis of magnetisation that accompanies mechanical oscillations of the particles is presented as well.

Acknowledgments
This work was supported by RFBR grants 16-31-00503, 17-41-590160, and 17-42-590504

6. Appendix:

Let us give a simple estimate for the reference times of the examined system. For this purpose, consider a solid spherical particle that moves slowly (in the Stokes regime) inside a viscoelastic medium. The resistance force acting on a particle of radius a in such a situation comprises the elastic and viscous parts:

$$f_{el} = -\gamma x = -6\pi a G x, \quad \text{and} \quad f_{visc} = -\zeta \dot{x} = -6\pi a \eta \dot{x}, \tag{10}$$

where G is the shear modulus and η is viscosity. Then the equation of motion of the particle of mass m writes as

$$m\ddot{x} = -\zeta \dot{x} - \gamma x. \tag{11}$$

As follows from (11), the time during which the particle velocity assumes its steady value (the inertial time) by the order of magnitude equals the ratio of the particle mass to the coefficient of viscous resistance of the medium, while the stress relaxation time is defined by the ratio of the viscous and elastic coefficients of the medium:

$$\tau_I \sim m/\zeta, \qquad \tau_R = \zeta/\gamma. \tag{12}$$

A typical stress relaxation time τ_R of a soft elastomer (weakly-linked silicone rubber) that makes the base of MRE is of the order of few seconds. With the assumed in above value of the shear modulus, this yields the elastomer viscosity to be in the interval from 1 and 10 kPa·s. Meanwhile, the inertial time τ_I for a particle of a few microns in diameter made of iron (specific density $\rho \sim 8 \times 10^3\,\mathrm{kg/m^3}$ is

$$\tau_I \simeq \rho V/(6\pi a \eta) = 2\rho a^2/9\eta \sim 10^{-12}\,s; \tag{13}$$

Here, to be particular, we set $a \simeq 5\,\mu$m and $\eta \simeq 10\,$kPa·s. Therefore the inertial time τ_I comes out to be more than ten orders of magnitude shorter than the stress relaxation time τ_R. This implies that when considering the processes, which the particle undergoes with the reference times about τ_R, the inertial forces may be safely neglected.

References

[1] Nikitin L V, Mironova L S, Stepanov G V and Samus A N 2001 *Polymer Science, series A* **43** 443–450
[2] Bellan C and Bossis G 2002 *International Journal of Modern Physics B* **16** 2447–2453
[3] Bose H 2007 *International Journal of Modern Physics B* **21** 4790
[4] Stepanov G V, Abramchuk S S, Grishin D A, Nikitin L V, Kramarenko E Yu and Khokhlov A R 2007 *Polymer* **48** 488–495
[5] Chertovich A, Stepanov G V, Kramarenko E Yu and Khokhlov A R 2010 *Macromolecular Materials and Engineering* **295** 336–341
[6] Becker T I, Raikher Yu L, Stolbov O V, Böhm V and Zimmermann K 2017 *Smart Materials and Structures* **26** 095035
[7] Kramarenko E Y, Chertovich A V, Stepanov G V, Semisalova A S, Makarova L A, Perov N S and Khokhlov A R 2015 *Smart Materials and Structures* **24** 035002
[8] Biller A M, Stolbov O V and Raikher Yu L 2014 *Journal of Applied Physics* **116** 114904
[9] Biller A M, Stolbov O V and Raikher Yu L 2015 *Physical Review E* **92** 023202
[10] Bozorth R M 1993 *Ferromagnetism* (Wiley-IEEE Press)

Influence of the geometry on magnetic interactions in a retina fixator based on a magnetoactive elastomer seal

T A Nadzharyan[1], L A Makarova[1], E G Kazimirova[1], N S Perov[1] and E Yu Kramarenko[1]

[1] Lomonosov Moscow State University, Faculty of Physics, 1-2 Leninskie Gory, Moscow, Russian Federation

E-mail: kram@polly.phys.msu.ru

Abstract. We study the effects the geometric configuration has on magnetic interactions between a magnetoactive elastomer (MAE) sample and various systems of permanent magnets for problems with both flat and curved geometry. MAEs consist of a silicone polymer matrix and iron filler microparticles embedded in it. Permanent magnets are cylindrical neodymium magnets arranged in a line on a flat or curved solid surfaces. We use computer simulations, namely the finite element method, in order to study the interaction force and magnetic pressure in a system with an MAE sample and permanent magnets. The model is based on classical Maxwell magnetostatics and two factors taking into account field dependence of MAE's magnetic properties and inhomogeneities caused by local demagnetization. We calculate magnetic pressure dependences on various geometric parameters of the system, namely, the diameter and the height of permanent magnets, the distance between the magnets and dimensions of MAE samples. This research aims to create a set of guidelines for choosing the geometric configuration of a retina fixator based on MAE seals to be used in eye surgery for retinal detachment treatment.

1. Introduction

Magnetoactive elastomers (MAEs) are polymer composite materials that have magnetically sensitive filler particles embedded in the polymer matrix [1-13]. The main property that characterizes these smart materials is their high responsiveness to external magnetic fields. Viscoelastic as well as magnetic properties of MAEs can be controlled externally [1-5]. The most widely used components of the composite are silicone-based soft matrices and micron-sized iron particles [3,4].

Magnetorheological effect, i.e. the changes to viscoelastic properties that occur in an MAE sample under the influence of external magnetic fields [1-5], and magnetodeformational effect, i.e. spontaneous deformation in magnetic field [6-11], serve as a basis for using MAEs as elements of tunable damping devices [12] and cell substrates [13]. This paper is dedicated to studying the MAEs for medical purposes, namely a surgical treatment of complicated retinal detachment cases [14-16]. It is proposed to seal the retinal break with an MAE while a system of permanent magnets placed outside of the eye provides the necessary magnetic field to turn the sample into a proper mechanical fixator. The MAE attracts to the system of magnets and also changes its properties to match the needs of the task at hand. The advantages of such approach include the flexibility of the material, ease of manufacturing and the lack of any eye trauma. The magnetic buckle consisting of several magnets

IOP Conf. Series: Journal of Physics: Conf. Series **994** (2018) 012002 doi:10.1088/1742-6596/994/1/012002

covered by medical silicone rubber is sutured to the sclera, and a MAE seal is introduced into the eye through a small incision and placed on the retina surface.

This paper continues and improves upon our previous combined theoretical and experimental study demonstrating high prospects of using an MAE as an element of a magnetic fixator for treatment of complicated retinal detachments [17]. Our previous work involved setting up the model and checking its validity through calculation of the interaction parameters for test flat configurations as well as comparison with experimental data [17]. Here we are exploring how varying the geometric configuration of the system affects the results. We choose system configurations with dimensions appropriate for eye surgery applications. Furthermore, we perform calculations for curved configurations corresponding to real eye problem and reveal how the fixator curvature affects its efficiency.

This paper is organized as follows. In the next section we describe the geometry of the problem. We then recreate this geometry in a modeling environment and discuss the modeling setup. After that we point out the main ideas and expressions required to solve the problem. The results and discussion are followed by conclusions.

2. Materials and geometry

In this paper we make use of two main geometric setups: a regular plane-like setup and a curved sphere-like setup. They are schematically shown in figure 1. We use cylindrical *NdFeB* magnets and a cuboid patch of an MAE sample that can then be deformed. Magnets are arranged in a periodic pattern and form a buckle. The sample is placed above the magnetic system. When it is not stated otherwise, we consider magnets with the radius $R = 1.5$ mm and the height $h_m = 1$ mm. This choice of the magnet dimensions is the result of eye surgery requirements. The diameter of the non-magnetic silicone buckle that is commonly used nowadays in scleral buckling for treatment of non-complicated retinal detachments does not exceed 5 mm [18,19]. In a magnetic fixator the permanent magnets are supposed to be placed inside such a silicon buckle, i.e. their size should be smaller than the buckle diameter.

In terms of magnetic moments $\vec{M}$ of the magnets, we consider two configurations: the first one corresponds to $\vec{M}$ vectors being aligned parallel to each other (figure 1a), and the second one corresponds to the anti-parallel alignment (figure 1b). One could expect magnetic field gradients in the anti-parallel configuration to be higher providing thus a higher force acting between a permanent magnet buckle and a MAE sample.

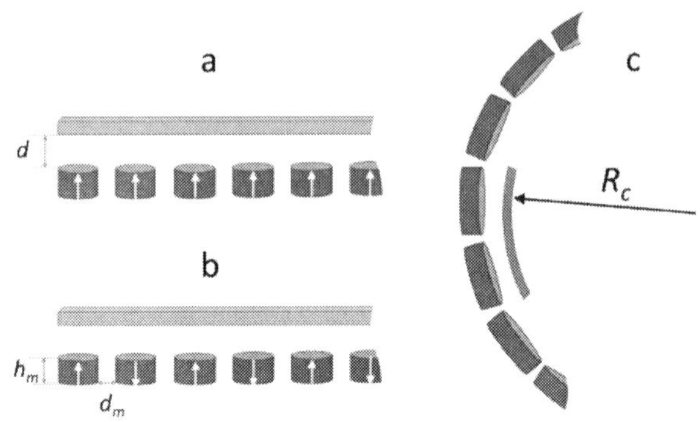

Figure 1. Schematic representation of magnetic systems under study. (a) a part of a system of ten magnets with parallel alignment of their magnetic moments and a MAE sample; (b) a part of a system of ten magnets with anti-parallel alignment of their magnetic moments and a MAE sample; (c) a part of a curved configuration of 22 cylindrical magnets and a deformed MAE sample placed on a spherical surface.

As for the MAE seal, we perform calculations for the sample containing 75mass % of iron particles. MAEs of this composition were used in our previous publication [17] in order to test the theoretical model. The MAE sample is cuboid-shaped and it either covers the entire upper side of the magnetic buckle or has the size equal to two periods of it. When it is not stated otherwise, the sample has the thickness $l_z = 0.2$ mm. The MAE sample of this size is designed to allow introducing the seal into an eye through a small scission of 0.6-0.8 mm.

3. Modeling setup

In our modeling of the interactions between a system of magnets and a MAE sample we use a continuous medium approach. Both the sample and the system of magnets are characterized by their shape and sizes as well as their relative coordinates. We think of them as magnetic materials that either produce or respond to the magnetic field. To calculate the magnetic field configuration as well as the interaction force Finite Element Method (FEM) was used. Commercially available software COMSOL Multiphysics was utilized to that end. The main work area was a cube with a size enough for boundary conditions' influence to become negligible. The sample and the system of magnets were placed according to symmetry. The *NdFeB* magnets were represented by cylinders with radii R, height h_m and magnetization in the following form: $\vec{M} = (0,0,M_z)^T$. The number of magnets in the system N and the distance between them d_m can also be controlled within the model.

We study two main geometric configurations: plane (figures 1a and 1b) and curved (figure 1c). The first one is much easier to consider, and as such most of the simulations are performed for the plane case. For this simulation the sample was considered to be a cuboid with set length l_x, width l_y and thickness l_z. The curved configuration corresponds to practical applications, that is, to both subsystems being placed along the curved surface of the human eye. For this simulation the sample is shaped by a spherical surface it is placed on and is characterized by the curvature radius R_c and the thickness l_z. The minimal distance between the faces of the magnets and the sample was denoted as d.

The conjugate gradient method was chosen as the main numerical algorithm for solving the magnetostatic problem. The relative tolerance was set to 10^{-6}. The mesh consisted of tetrahedral elements with second order Lagrange polynomials. The maximum element size was chosen for each subdomain after performing several consecutive manual refinements of the mesh to achieve the relative deviation of less than 0.1% for the solution parameters on subsequent meshing steps. The mesh was also refined additionally between the sample and the system of magnets as well as in their close proximity to improve the accuracy.

For each configuration distributions of magnetic energy density W_m, magnetic field $\vec{H}$ and stress $\vec{n}\hat{T}$ as well as the electromagnetic force $\vec{F}_{em}$ acting on the sample were calculated. Effective pressure P is obtained as total force $\vec{F}_{em}$ per area S: $P = \left|\vec{F}_{em}\right| / S$.

The comparison of the modeling results with the experimental data was performed in our previous work [17]. It has shown that two factors need to be considered when modeling the interaction between a MAE sample and a system of magnets. The first one is the magnetic field dependence of the MAE magnetic permeability function $\mu(\vec{H})$. It is one of the main properties of magnetoactive materials, so it is only natural to factor it in. The second one is demagnetization and form-factor [20]. It is known that the magnetic field inside a real object does not match external magnetic field even with permeability taken into account. It is because of how the shape of the object (or its boundary) affects the distribution of the field inside. It can be illustrated by considering the field boundary condition for the object in question and introducing a distribution of effective "magnetic charges" to satisfy the aforementioned boundary condition.

These two factors combined allow us to obtain the magnetization distribution inside the magnets which is not uniform and the dependence of the sample's magnetic permeability on the magnetic field inside the sample instead of just an external magnetic field from the source. For the field dependence itself we use the Langevin model and do not introduce magnetic hysteresis into our model, since the experimental results for our samples showed hysteresis to be insignificant. Our previous work contains a more detailed explanation of the modeling process with both factors taken into account as well as information about experimental support as it outlines the basis of this particular research [17]. Experimental data for cylindrical neodymium magnets (R = 2 mm and h_m = 1 mm) as well as MAE samples with 75 mass percentage of magnetic filler obtained in [17] was used to derive the necessary uniqueness conditions for the magnetization distribution and to calculate the parameters of the Langevin model respectively. These results could be found in our previous publication [14].

4. Magnetic field problem and force calculations

We calculate the force resulting from stress using the standard procedure [20]:

$$\vec{F} = \oint_S \vec{n}\hat{T}_{ext} dS$$

Here $\vec{n}$ is the external normal vector for the object and $\hat{T}_{ext}$ is the Cauchy stress tensor outside of the object. If the problem in question is purely electromagnetic, then $\hat{T}_{ext}$ stands for the Maxwell stress tensor in the corresponding area.

For air on the sample boundary the stress tensor is:

$$\hat{T}_{ext} = -p\hat{I} - \frac{1}{2}\left(\vec{H}\cdot\vec{B}\right)\hat{I} + \vec{H}\otimes\vec{B}$$

Here p is pressure and $\hat{I}$ is an identity tensor.

The magnetic field problem is static, so we have the following field equations:

$$\begin{cases} \vec{\nabla}\times\vec{H} = \vec{j}(\vec{r}) = 0 \\ \vec{\nabla}\cdot\vec{B} = 0 \end{cases}$$

Magnetic field configuration strongly impacts both the pressure distribution on the sample surface and the overall interaction force. This is due to these characteristics being mainly affected by field inhomogeneities. Magnetic field structure plays an even larger role, if one considers small distances which are relevant to the practical medical applications of the present study.

5. Results and discussion

The interaction force (and the resulting mean pressure) depends on the geometric configuration of the system. Varying geometric parameters allows to find an optimal configuration as well as to study the sensitivity of the system to changes in the geometry. Figures 2-7 show dependences of the interaction force (pressure) on various geometric parameters of both the magnetic buckle and the MAE sample calculated for the flat configuration. The distance between the closest faces of the magnetic buckle and the MAE sample was fixed at 1.3 mm which roughly corresponds to the one in a real system (the thickness of the human eye sclera together with retina is approximately equal to 0.7 mm and a silicon layer on the surface of magnets in a buckle gives additional 0.6 mm of the distance).

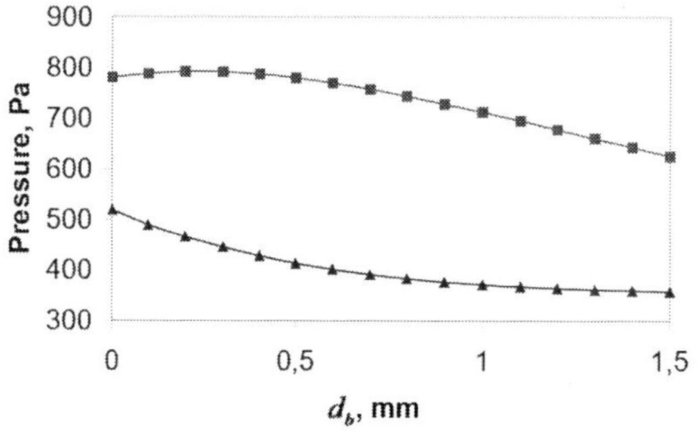

Figure 2. Pressure dependence on the distance between the magnets for $R = 1.5$ mm and $h_m = 1$ mm. The MAE sample parameters are as follows: l_x changes with d_b, $l_y = 3$ mm and $l_z = 0.2$ mm. The distance between the system of the magnets and the sample was kept at 1.3 mm. Line with triangle markers corresponds to the parallel magnetization alignment and line with square markers corresponds to the anti-parallel alignment.

In figure 2 we plot the pressure dependence on the distance d_b between the magnets in the magnetic buckle. In these calculations, the length of the MAE sample was also increased simultaneously with increasing length of the magnetic buckle as a whole with d_b. In this case the interaction force, from one hand, should grow due to the larger area S of the sample. However, distancing magnets from each other also makes the structure of the magnetic field more uniform, which in turn decreases the resulting force. As a result, the pressure slightly decreases with d_b. For anti-parallel configuration the inhomogeneity is much more prevalent, which leads to the pressure being practically constant in the range of small distances d_b up to 0.5 mm. Figure 2 clearly confirms that the pressure produced by the magnetic buckle with alternating magnetization orientation is indeed higher. The pressure difference for two configurations (figures 1a and 1b) almost reaches 100%.

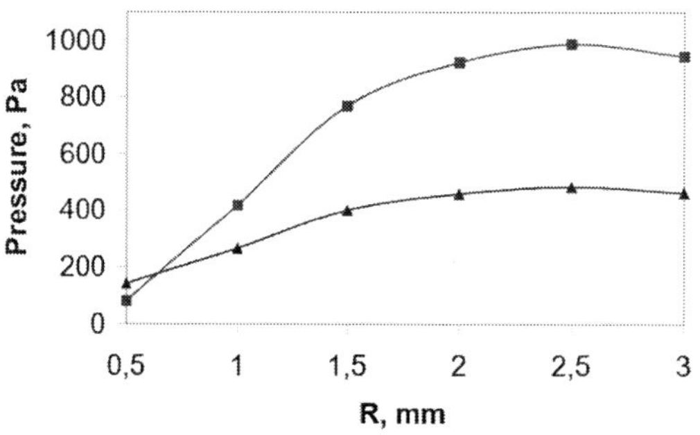

Figure 3. Pressure dependence on the size (radii) of the magnets for $h_m = 1$ mm and $d_b = 0.6$ mm. Sample parameters are as follows: l_x and l_y change with R and $l_z = 0.2$ mm. The distance between the system of the magnets and the sample was kept at 1.3 mm. Line with triangle markers corresponds to the parallel magnetization alignment and line with square markers corresponds to the anti-parallel alignment.

Figure 3 shows the pressure dependence on the permanent magnet radius R. The corresponding graphs were obtained assuming the overall magnetic moment of each magnet to be changing slowly with its radius. The MAE sample's width and length were increased together with R. One can see that in case of small magnets, namely when the magnet radius is smaller than the distance between the magnets, the pressure produced in the anti-parallel magnetization configuration is even lower than in the co-directional magnetization system due to a small effective magnetic field in between the magnets. The pressure expectedly increases with the magnet radius for the both permanent magnet configurations and for R exceeding d_b it becomes higher for the anti-parallel system.

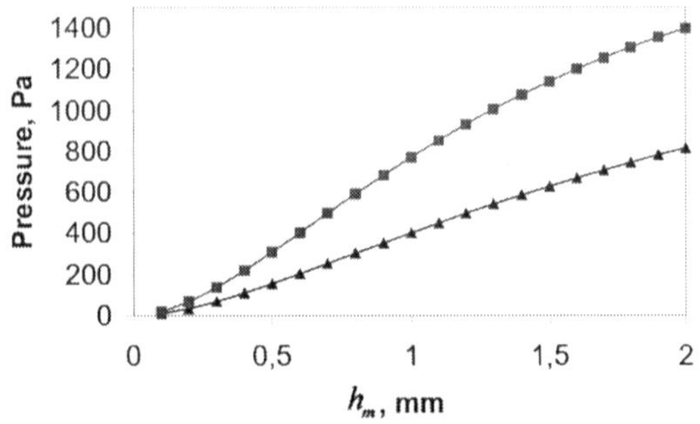

Figure 4. Pressure dependence on the height of the magnets for R = 1.5 mm and d_b = 0.6 mm. MAE sample parameters are as follows: l_x = 35.4 mm, l_y = 3 mm and l_z = 0.2 mm. The distance between the system of magnets and the sample was kept at 1.3 mm. Line with triangle markers corresponds to the parallel alignment of magnet magnetic moments and line with square markers corresponds to the anti-parallel alignment.

Increasing amount of magnetic material with a growing height of the magnets in the buckle obviously causes further increase in the pressure (see figure 4). Approximately two-fold difference in pressures for two types of magnetization orientation within the buckle is realized in the whole range of h_m.

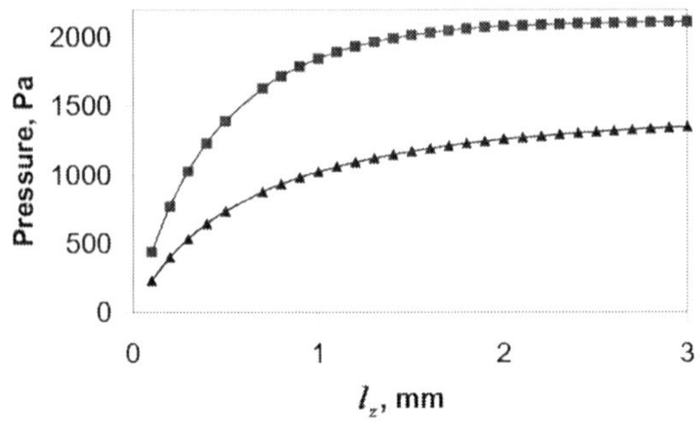

Figure 5. Pressure dependence on the MAE sample thickness for R = 1.5 mm, d_b = 0.6 mm and h_m = 1 mm. Sample parameters are as follows: l_x = 35.4 mm and l_y = 3 mm. The distance between the system of magnets and the sample was kept at 1.3 mm. Line with triangle markers corresponds to the parallel alignment of magnet magnetic moments and line with square markers corresponds to the anti-parallel alignment.

Figure 5 demonstrates the pressure dependence on the thickness of the MAE sample at fixed dimensions of the magnetic buckle. A rapid increase of the pressure with l_z is observed at MAE thicknesses smaller than 1 mm while at higher values a kind of a saturation takes place because the "additions" the magnets are located further and further away from the sample, and the magnetic field decreases rapidly with distance.

The obtained results presented in figures 2-5 demonstrate that for system dimensions appropriate for eye surgery purposes and especially at small distances d changing geometric parameters of the system can lead to a substantial change in interaction forces. This is the most apparent in the cases of the magnet radii (figure 3), the height (figure 4) and the sample thickness (figure 5). All the dependences are non-linear and exhibit linear-like behavior for small values of the geometric parameters. This can be explained by the field influence weakening with larger distances.

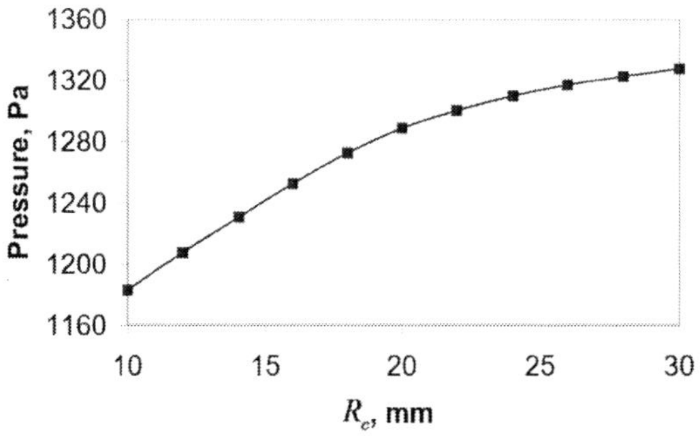

Figure 6. Pressure dependence on the curvature of the system. Parameters of the system of magnets are: R = 1.5 mm, d_b = 0.4 mm on average and h_m = 1 mm. Parameters of the undeformed sample: l_x = 6.8 mm, l_y = 2 mm and l_z = 0.2 mm. The real l_x and l_y change with R_c. The reference distance between the system of magnets and the sample d = 0.7 mm.

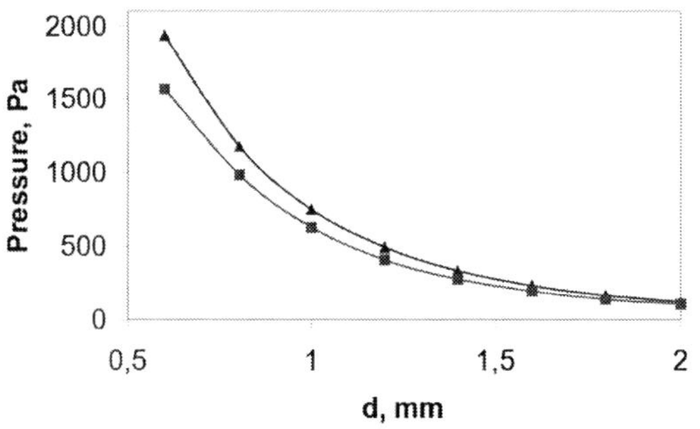

Figure 7. Pressure dependence on the distance between the sample and the system of magnets for R = 1.5 mm, d_b = 0.4 mm and h_m = 1 mm. The undeformed sample parameters are as follows: l_x = 6.8 mm, l_y = 2 mm and l_z = 0.2 mm. Line with triangle markers corresponds to the flat case and line with square markers corresponds to the curved case. The curvature radius for the curved case: R_c = 12 mm.

As the next step we investigate the effect of the magnetic system curvature on the magnetic interactions. In figure 6 the pressure vs the curvature radius is presented at fixed values of all other geometrical parameters of the magnetic fixator while in figure 7 dependence of the pressure on the distance between the magnetic buckle and the MAE sample is shown for the flat geometry and for the curved one with the curvature radius equal to that of a human eye.

One can see that increasing R_c causes a slight increase in the pressure. For high values of R_c the curvature influence on the pressure is small, but for R_c around 12 mm, which corresponds to the curvature of a real human eye, the effect is more noticeable. The difference between the plane-like configuration and the sphere-like one with R_c = 12 mm is about 19%. For the R_c of "infinity", or the flat case, we have the pressure of roughly 1432 Pa. As such, while the flat geometry can be used to determine the tendencies and approximate results, it is imperative to study the curved geometry to obtain more accurate results for medical purposes.

The developed theoretical approach allows to calculate not only the average magnetic forces and pressure in the system but also the distribution of magnetic field within the magnetic fixator outside the permanent magnets (figure 8). One can see that the magnetic field concentrates near the magnetic buckle rapidly decreasing with the distance. It is also evident that the magnetic field inside the sample is inhomogeneous and is smaller than the outside field due to demagnetization. The structure of the field is substantially different from that of a flat case, since magnets are solid objects and can't form a perfect curved surface.

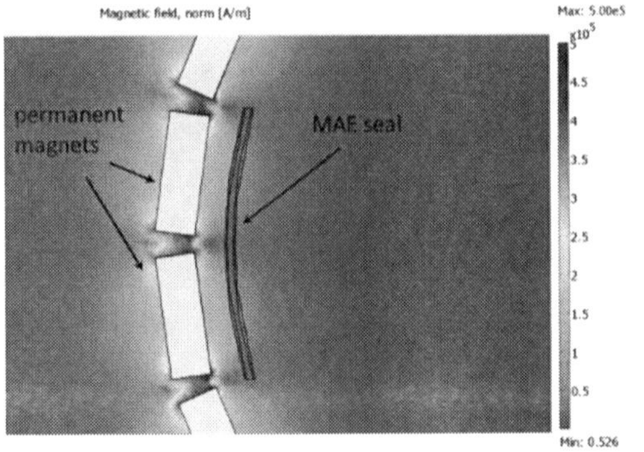

Figure 8. Magnetic field configuration for the curved geometry. Parameters of the system of magnets are: R = 1.5 mm, d_b = 0.4 mm on average and h_m = 1 mm. Parameters of the undeformed sample: l_x = 6.8 mm, l_y = 2 mm and l_z = 0.2 mm. The curvature radius R_c = 12 mm.

6. Conclusions

In this work we continue studying the interaction between a system of permanent magnets and an MAE sample. This study in particular was aimed at simulating various possible geometric configurations of the system and considering the effect that changes in the geometry have on the interaction parameters. We used the finite element modeling in the commercially available software COMSOL Multiphysics to simulate two principal geometries of the problem: flat and curved cases. The flat case is used as a reference due to the ease of its analysis, and the curved case is used to study the specifics of the problem that would be present in the real scenario of medical applications of an MAE-based retina fixator.

It has been demonstrated that within the range of dimensions suitable for eye surgery applications even slight changes in geometry can strongly affect the interaction force and magnetic pressure. As such, it is important to both be precise when manufacturing the components of the retina fixator device and consult the modeling results to choose the optimal configuration for the task at hand.

The effective pressure dependences on the size of the magnets, the distance between them, their magnetic configuration, the size of the sample and the type of the system geometry have been calculated. The present results further reinforce the idea of using MAE-based retina fixators for surgical treatment of retinal detachment that we proposed in our previous work.

Acknowledgments
Financial support of the Russian Science Foundation (grant no. 16-15-00208) is gratefully acknowledged.

References

[1] Jolly M R, Carlson J D, Munoz B C and Bullions T A 1996 The magnetoviscoelastic response of elastomer composites consisting of ferrous particles embedded in a polymer matrix *J. Intell. Mater. Syst. Struct.* 7 613-22
[2] Filipcsei G, Csetneki I, Szilgyi A and Zrinyi M 2007 Magnetic field-responsive smart polymer composites *Adv. Polym. Sci.* **206** 137-89
[3] Ubaidillah, Sutrisno J, Purwanto A and Mazlan S A 2015 Recent progress on magnetorheological solids: materials, fabrication, testing, and applications *Adv. Eng. Mater.* **17** 563-97
[4] Sorokin V V, Stepanov G V, Shamonin M, Monkman G J, Khokhlov A R and Kramarenko E Yu 2015 Hysteresis of the viscoelastic properties and the normal force in magnetically and mechanically soft magnetoactive elastomers: effects of filler composition, strain amplitude and magnetic field *Polymer* **76** 191-202
[5] Kramarenko E Yu, Chertovich A V, Stepanov G V, Semisalova A S, Makarova L A, Perov N S and Khokhlov A R 2015 Magnetic and viscoelastic response of elastomers with hard magnetic filler *Smart Mater. Struct.* **24** 035002
[6] Zrinyi M, Barsi L and Buki A 1996 Deformation of ferrogels induced by nonuniform magnetic fields *J. Chem. Phys.* **104** 8750-6
[7] Abramchuk S, Kramarenko E, Stepanov G, Nikitin L V, Filipcsei G, Khokhlov A R and Zrinyi M 2007 Novel highly elastic magnetic materials for dampers and seals I.: preparation and characterization of the elastic materials *Polym. Adv. Technol.* **18(11)** 883-90
[8] Abramchuk S, Kramarenko E, Grishin D, Stepanov G, Nikitin L V, Filipcsei G and Zrinyi M 2007 Novel highly elastic magnetic materials for dampers and seals II.: material behaviour in a magnetic field *Polym. Adv. Technol.* **18(7)** 513-8
[9] Stepanov G V, Chertovich A V and Kramarenko E Yu 2012 Magnetorheological and deformation properties of magnetically controlled elastomer with hard magnetic filler *JMMM* **324(21)** 3448-51
[10] Stepanov G V, Kramarenko E Yu and Semerenko D A 2013 Magnetodeformational effect of the magnetoactive elastomers and its possible applications *J. Phys. Conf. Ser.* **412** 012031
[11] Ginder J M, Clark S M and Schlotter W F 2002 Magnetostrictive phenomena in magnetorheological elastomers *Int. J. Mod. Phys. B* **16** 2412-8
[12] Li Y, Li J, Li W and Du H 2014 A state-of-the-art review on magnetorheological elastomer devices *Smart Mater. Struct.* **23** 123001
[13] Mayer M, Rabindranath R, Boerner J, Hoerner E, Bentz A, Salgado J, Han H, Boese H, Probst J, Shamonin M, Monkman G J and Schlunck G 2013 Ultra-soft PDMS-based magnetoactive elastomers as dynamic cell culture substrata *PLoS ONE* **8(10)** e76196
[14] Zhang Z, Liang X, Sun D and Peng S 2011 The scleral buckling of primary rhegmatogenous retinal detachment under the surgical microscope *Ophthal. Surg. Las. Im.* **42(2)** 96-101
[15] Schaal S, Sherman M P, Barr C C and Kaplan H J 2011 Primary retinal detachment repair: comparison of 1-year outcomes of four surgical techniques *Retina* **31(8)** 1500–4
[16] Mehta S, Blinder K J, Shah G K and Grand M G 2011 Pars plana vitrectomy versus combined pars plana vitrectomy and scleral buckle for primary repair of rhegmatogenous retinal detachment *Can. J. Ophthalmol.* **46(3)** 237–41
[17] Makarova L A, Nadzharyan T A, Alekhina Yu A, Stepanov G V, Kazimirova E G, Perov N S and Kramarenko E Yu 2017 Magnetoactive elastomer as an element of a magnetic retina fixator *Smart Mater. Struct.* **26** 095054
[18] Storey P, Alshareef R, Khuthaila M, London N, Leiby B, Kaiser R and Wills PVR Study Group

2014 Pars plana vitrectomy and scleral buckle versus pars plana vitrectomy alone for patients with rhegmatogenous retinal detachment at high risk for proliferative vitreoretinopathy *Retina* **34(10)** 1945-51

[19] Rush R B, Simunovic M P, Sheth S, Kratz A and Hunyor A P 2013 Pars plana vitrectomy versus combined pars plana vitrectomy-scleral buckle for secondary repair of retinal detachment *Ophthal. Surg. Las. Im. Retina* **44** 374-9

[20] Chikazumi S 1997 *Physics of Ferromagnetism*, ed C D Graham Jr (Oxford: Oxford Science Publication)

Preparation, structure and magnetic properties of synthetic ferrihydrite nanoparticles

S V Stolyar[1,2], R N Yaroslavtsev[1,2], O A Bayukov[2], D A Balaev[1,2], A A Krasikov[2], R S Iskhakov[2], A M Vorotynov[2], V P Ladygina[3], K V Purtov[4], M N Volochaev[2]

[1]Siberian Federal University, Krasnoyarsk, 660041, Russia
[2]Kirensky Institute of Physics, SB RAS, Krasnoyarsk, 660036, Russia
[3]International scientific centre for organism extreme states research attached Presidium of KSC, SB RAS, Krasnoyarsk, 660036, Russia
[4]Institute of Biophysics, SB RAS, Krasnoyarsk, 660036, Russia

rauf@iph.krasn.ru

Abstract. Superparamagnetic ferrihydrite powders with average nanoparticle sizes of 2.5 nm produced by the chemical deposition method. Static and dynamic magnetic properties are measured. As a result of ultrasonic treatment in the cavitation regime of suspensions of ferrihydrite powders in a solution of the albumin protein, the Fe ions are reduced to the metallic state. A sol of ferrihydrite nanoparticles is prepared in an aqueous solution of arabinogalactan polysaccharide.

1. Introduction

Fine powders (powders with particle sizes less than 10 nm) based on oxides, hydroxides, and oxyhydroxides of iron are the object of increased interest at the present time.This circumstance is due to the practical use of powders in such areas as catalysis, biomedicine, etc. [1]. Bulk ferrihydrite is an antiferromagnet. However, as the particle sizes decrease to nanoscale, the magnetic properties of the powder change radically. Nanoparticles of ferrihydrite acquire a constant magnetic moment due to incomplete compensation of the magnetic moments of the sublattices. Another important practical feature of ferrihydrite is its absolute biocompatibility, because this mineral is the core of the ferritin protein complex, which is the main carrier of iron in all higher living organisms. Therefore, ferrihydrite powders can compete with powders of conventional ferro- and ferrimagnets used in various practical applications [2], including for targeted drug delivery, as well as for contrasting in magnetic resonance imaging [3].

A method for obtaining stable sols of ferrihydrite nanoparticles of biogenic origin was developed in [4] (ferrihydrite nanoparticles are formed on the cell surface during the cultivation of Klebsiella oxytoca bacteria). As it turned out, the magnetic properties of biogenic nanoparticles of ferrihydrite can be modified by heat treatment [5], however, it is not possible to produce a stable sol on the basis of annealed powders. As a matrix for the preparation of stable sols based on chemically produced ferrihydrite powders, it is possible to use organic preparations such as water-soluble proteins albumin or natural polysaccharide arabinogalactan.

The aim of this work was the production of ferrihydrite powders and finding ways to preparation of sols based on them. The process of preparation of ferrihydrite-based sols consists in the ultrasonic treatment of suspensions of nanoparticles.

2. Experimental methods

Synthetic ferrihydrite nanoparticles were prepared chemically [6]. At room temperature and constant stirring, a 1 M alkali solution of NaOH was added to a 0.2 M solution of iron nitrate $Fe(NO_3)_3 \cdot 9H_2O$ to a neutral pH value. The precipitate formed was collected on a filter. Then, the precipitate was washed and dried at room temperature. The electron microscopy examination was carried out on a Hitachi HT7700 transmission electron microscope. The Mössbauer spectra were measured on an MS-1104Em spectrometer with a $^{57}Co(Cr)$ source. The chemical isomer shifts were given with respect to α-Fe. The magnetic measurements were carried out on a vibrating-sample magnetometer [7]. The temperature dependences of the magnetic moment M(T) were measured both in the mode of cooling without a magnetic field (zero field cooling (ZFC)) and in the mode of cooling in an external magnetic field (field cooling (FC)). The magnetic resonance spectra were recorded on a Bruker ELEXSYS 560 spectrometer operating in the X band mode (the characteristic microwave radiation frequency was ~9.4 GHz) at temperatures in the range from 100 to 300 K. Ultrasonic treatment was performed on the Volna UZTA-0.4/22-OM apparatus. The intensity of ultrasonic treatment was over 10 $W \cdot cm^{-2}$ and the frequency was 22 kHz. The treatment time was 4–24 min. Ultrasonic treatment was carried out in an aqueous medium, in a solution of protein albumin and arabinogalactan.

3. Results and discussion

Figure 1 shows the results of transmission electron microscopy of the obtained ferrihydrite nanoparticles. Figure 1 (a) shows the microdiffraction pattern of nanoparticles, which is characteristic of ferrihydrite nanoparticles [8,9]. Two diffuse reflections with interplanar distances d_1 = 1.6 Å, d_2 = 2.7 Å are observed. The average particle size was D = 2.5 nm (Fig. 1 (b)).

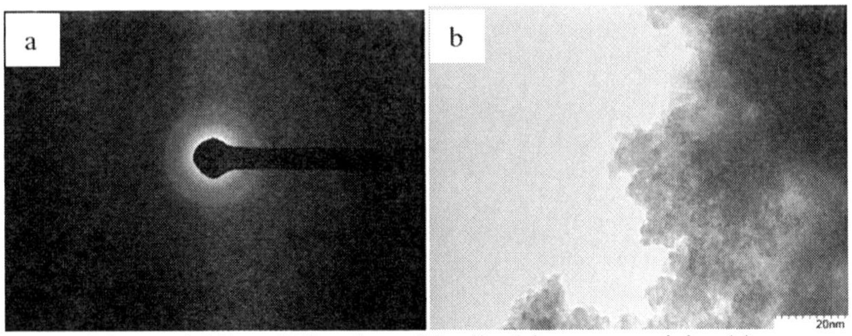

Figure 1. Microdiffraction of ferrihydrite nanoparticles (a), transmission electron microscope images (b).

The temperature dependences of the magnetization, measured in the ZFC and FC modes, shown in Fig. 2a, exhibit a characteristic superparamagnetic (SP) behavior. At $T < T_B$, the *M(H)* curves exhibit a hysteresis, see the inset of the fig.2.a Analysis of the *M(H)* dependences at temperatures $T > T_B$ (see the fig.2b) was carried out using the standard approach for systems of noninteracting antiferromagnetic nanoparticles. In this approach, the magnetic moment of the sample is determined by the SP behavior of individual particles, taking into account their distribution over magnetic moments, and also the $\chi_{AF} \cdot H$ component, which determines the antiferromagnetic contribution to the magnetization. In this case, the *M(H)* dependences are described by the following expression [7,10,11]:

$$M(H) = N_P \int_{\mu_{min}}^{\mu_{max}} L(\mu_P, H) f(\mu_P)\, \mu_P d\mu_P + \chi_{AF} \cdot H. \quad (1)$$

In this expression $L(\mu_P, H)$ – Langevin function, $f(\mu_P)$ – distribution function of the magnetic moment of the particles μ_P, N_P – number of particles per unit weight of the sample. The lognormal distribution was used [10]. From the temperature dependence $\langle\mu_P\rangle(T)$, by extrapolation to $T = 0$ K, values of $\langle\mu_P\rangle(T = 0)$ were determined, which amounted to 173 μ_B. The fig.2b shows experimental isotherms of magnetization $M(H)$ as well as the results of the best fit using eq.(1) and partial contributions of the first (superparamagnetic) and the second (antiferromagnetic $\chi{\cdot}H$) terms of eq.(1).

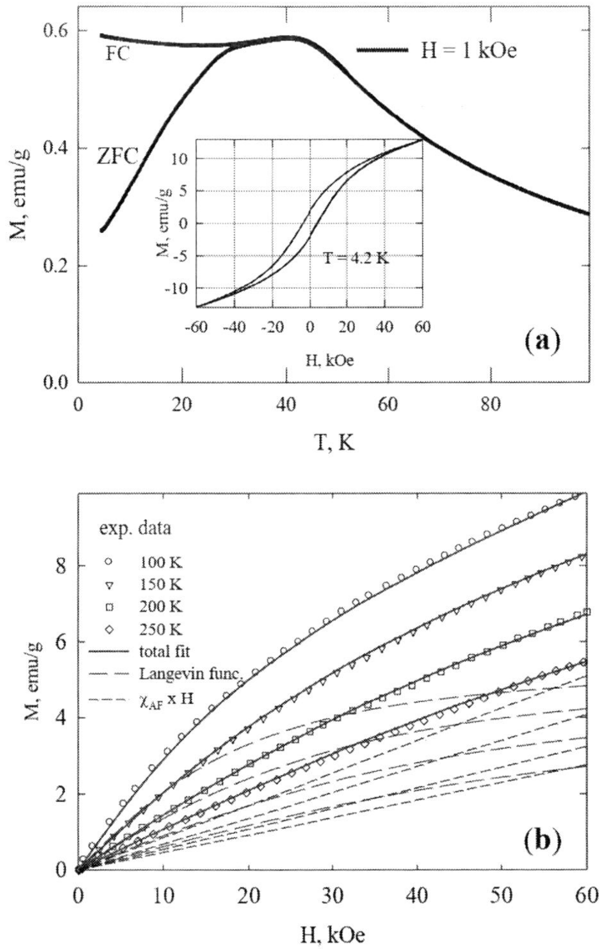

Figure 2. Magnetic properties of synthetic ferrihydrite nanoparticles. (a) The temperature dependences of the magnetic moment M(T) measured in the ZFC and FC modes. Inset in fig.2(a) – hysteresis dependence $M(H)$ at $T = 4.2$ K. (b) –Magnetization curves $M(H)$ obtained at different temperatures and the results of the best fit using eq.(1). Partial contributions of superparamagnetic antiferromagnetic $\chi{\cdot}H$ terms of eq.(1) are also shown.

Table 1 shows the values of the particle size D, the blocking temperature T_B, the coercive field H_C, the saturation magnetization M_S and the antiferromagnetic susceptibility χ_{AF}. For comparison, the table also shows the characteristics of ferrihydrite nanoparticles obtained as a result of cultivation of bacteria [12].

Table 1. Characteristics of the ferrihydrite samples

Sample	D, nm	T_B, K	H_C, kOe	M_s, Gs	χ_{AF}, 10^{-4} emu/(g·Oe)
Biogenic [5,7]	2	23.3	3.6	26	0.6
Synthetic	2.5	40	3.8	25	1.1

Ferromagnetic resonance spectra of ferrihydrite nanoparticles are symmetrical curves of the Lorentz shape, and the absorption maximum is located near ω/γ value, which for a given frequency is 3350 Oe. For ferrihydrite nanoparticles in the investigated temperature range, the intensity of the FMR signal decreased linearly, indicating that the nanoparticles are in the unlocked, SP state.

According to the Raikher-Stepanov theory [13,14], in the powders of randomly oriented particles of ferromagnets and ferrites, the width of the absorption line is a nonmonotonic function of temperature:

$$\Delta \mathrm{H(T)} = \Delta H_S(T) + \Delta H_U(T), \quad (2)$$

where $\Delta H_S(T)$ – contribution to the broadening due to superparamagnetism of nanoparticles, $\Delta H_U(T)$ – contribution to the broadening due to the scatter in the directions of the anisotropy fields of the particles (inhomogeneous broadening), which is decisive at low temperatures.

Figure 3 shows the values of the width of the FMR line as a function of temperature and the result of fitting of experimental data using the Raikher-Stepanov theory. The curve shown in Figure 3 is characterized by two adjustable parameters: *KV* and *MV*. Table 2 shows the values obtained, as well as *KV* and *MV* of nanoparticles of ferrihydrite of biogenic origin. The table also shows the corresponding parameters of ferritin complexes calculated from the temperature curves of FMR [15].

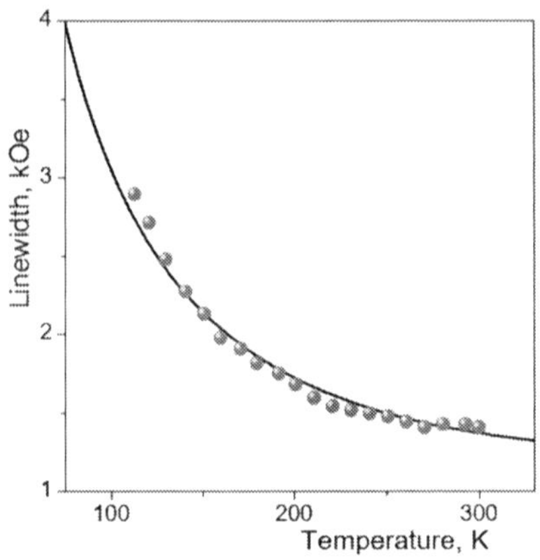

Figure 3. Temperature dependences of the width of the ferromagnetic resonance line for ferrihydrite nanoparticles. The solid line is the result of the fit.

Table 2. Fitting parameters of $\Delta H(T)$curves

Sample	*KV*, erg	*MV*, emu
Biogenic	$1.25 \cdot 10^{-14}$	$2.37 \cdot 10^{-18}$
Synthetic	$2 \cdot 10^{-14}$	$2.4 \cdot 10^{-18}$
Ferritin[15]	$2.5 \cdot 10^{-14}$	$1.9 \cdot 10^{-17}$

Figure 4 (curve 1) shows the Mossbauer spectrum of chemically prepared ferrihydrite nanoparticles, characterized by a paramagnetic doublet. Table 3 shows the results of the interpretatation. Three main non-equivalent positions of Fe^{3+} ions having octahedral coordination are registered: positions Fe1 and Fe2 with a relatively small degree of distortion of local symmetry and position of Fe3 with a large degree of distortion. After ultrasonic treatment of nanoparticles in water (curve 2), no significant changes in Mossbauer parameters were observed except for some redistribution in octahedral positions of Fe. Figure 4 also shows the Mössbauer spectrum of nanoparticles after ultrasonic treatment in an albumin solution (curve 3). In this case, the spectrum contains an additional sextet. The parameters of the sextet coincide with the Mössbauer parameters of the bcc phase of iron. Thus, ultrasonic treatment in protein solution leads to the reduction of Fe ions to the metallic state (19%).

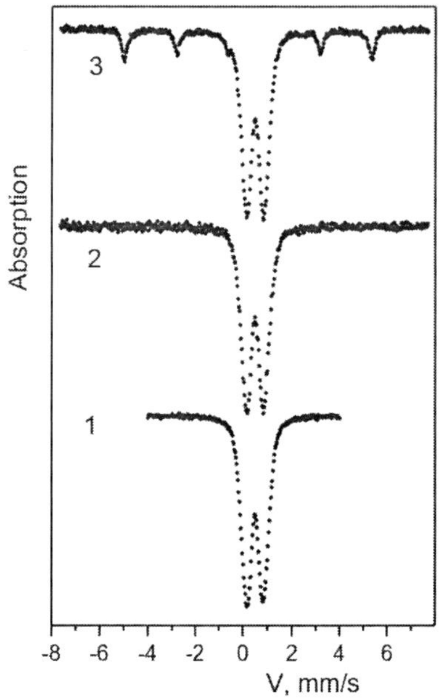

Figure 4. Mossbauer spectra. Initial nanoparticles - 1. After ultrasound treatment in water - 2 and in albumin solution -3.

To obtain the nanoparticle-arabinogalactan complexes, 1% sol of the ferrihydrite particles in distilled water was used. A solution of arabinogalactan (1%) in distilled water was poured onto the sol of the particles in a ratio of 1:1.Homogenization was performed by ultrasound for 10 min using an ultrasonic radiator -UltrasonicdisintegratorUD-20 (Techpan, Poland). The resulting complexes were washed from the excess unbound polysaccharide by centrifuging the particle sol at 35,000 g on a centrifuge Avanti® J-E (Beck-man Coulter, USA). The resulting precipitate was collected and re-homogenized in distilled water. The cycle was repeated 3 times. The final result was a stable sol of the complexes in distilled water, which was used for further research. The resulting complex has the property of restoring, that is, the dried sol particles with the addition of water also form a sol. The possibility of making a sol on the basis of synthetic nanoparticles of ferrihydrite in a solution of albumin was also investigated. The fraction of particles transferred to the sol was insignificant.

Table 3. Mossbauer parameters of ferrihydrite. IS is the isomeric chemical shift, QS is the quadrupole splitting, W is the width of the absorption line, H is the hyperfine field at the iron core, and A is the share population of the position.

	IS, mm/s	H, kOe	QS, mm/s	W, mm/s	A	Position
Initial nanoparticles	0.351		0.51	0.35	0.53	Fe1
	0.355		0.86	0.31	0.37	Fe2
	0.359		1.21	0.26	0.10	Fe3
After ultrasonic in water	0.349		0.49	0.33	0.42	Fe1
	0.352		0.79	0.30	0.38	Fe2
	0.355		1.16	0.33	0.20	Fe3
After ultrasonic in albumin solution	0.006	332	0	0.25	0.19	**α-Fe**
	0.350		0.53	0.35	0.35	Fe1
	0.351		0.84	0.36	0.37	Fe2
	0.357		1.22	0.30	0.10	Fe3

4. Conclusions

Ferrihydrite powders are produced by the chemical deposition method. Static and dynamic magnetic properties are measured. A sol of ferrihydrite nanoparticles is made in an aqueous solution of a natural polysaccharide, arabinogalactan.

Acknowledgments

The reported study was funded by Russian Foundation for Basic Research and Krasnoyarsk region according to the research project № 17-43-240527 and Russian Foundation for Basic Research project № 16-03-00969. Support by the Special Program for Siberian Federal University of the Ministry of Education and Science of the Russian Federation is acknowledged.

References

[1] Tartaj P, Morales M a del P, Veintemillas-Verdaguer S, Gonz lez-Carre o T and Serna C J 2003 The preparation of magnetic nanoparticles for applications in biomedicine *J. Phys. D. Appl. Phys.* **36** R182–97

[2] Dobretsov K, Stolyar S and Lopatin A 2015 Magnetic nanoparticles: a new tool for antibiotic delivery to sinonasal tissues. Results of preliminary studies. *Acta Otorhinolaryngol. Ital. organo Uff. della Soc. Ital. di Otorinolaringol. e Chir. Cerv.-facc.* **35** 97–102

[3] Inzhevatkin E V., Morozov E V., Khilazheva E D, Ladygina V P, Stolyar S V. and Falaleev O V. 2015 Elimination of Iron-Containing Magnetic Nanoparticles from the Site of Injection in Mice: a Magnetic-Resonance Imaging Study *Bull. Exp. Biol. Med.* **158** 807–11

[4] Stolyar S V, Bayukov O A, Balaev D A, Iskhakov R S, Ishchenko L A, Ladygina V P and Yaroslavtsev R N 2015 Production and magnetic properties of biogenic ferrihydrite nanoparticles *J. Optoelectron. Adv. Mater.* **17** 968–72

[5] Balaev D A, Krasikov A A, Dubrovskii A A, Semenov S V., Bayukov O A, Stolyar S V., Iskhakov R S, Ladygina V P and Ishchenko L A 2014 Magnetic properties and the mechanism of formation of the uncompensated magnetic moment of antiferromagnetic ferrihydrite nanoparticles of a bacterial origin *J. Exp. Theor. Phys.* **119** 479–87

[6] Michel F M, Ehm L, Antao S M, Lee P L, Chupas P J, Liu G, Strongin D R, Schoonen M A A, Phillips B L and Parise J B 2007 The Structure of Ferrihydrite, a Nanocrystalline Material *Science (80-.).* **316** 1726–9

[7] Balaev D A, Krasikov A A, Dubrovskiy A A, Popkov S I, Stolyar S V, Bayukov O A, Iskhakov R S, Ladygina V P and Yaroslavtsev R N 2016 Magnetic properties of heat treated bacterial ferrihydrite nanoparticles *J. Magn. Magn. Mater.* **410** 171–80

[8] Kukkadapu R K, Zachara J M, Fredrickson J K, Smith S C, Dohnalkova A C and Russell C K 2003 Transformation of 2-line ferrihydrite to 6-line ferrihydrite under oxic and anoxic conditions *Am. Mineral.* **88** 1903–14

[9] Guyodo Y, Banerjee S K, Lee Penn R, Burleson D, Berquo T S, Seda T and Solheid P 2006 Magnetic properties of synthetic six-line ferrihydrite nanoparticles *Phys. Earth Planet. Inter.* **154** 222–33

[10] Silva N J O, Amaral V S and Carlos L D 2005 Relevance of magnetic moment distribution and scaling law methods to study the magnetic behavior of antiferromagnetic nanoparticles: Application to ferritin *Phys. Rev. B* **71** 184408

[11] Gilles C, Bonville P, Rakoto H, Broto J M, Wong K K W and Mann S 2002 Magnetic hysteresis and superantiferromagnetism in ferritin nanoparticles *J. Magn. Magn. Mater.* **241** 430–40

[12] Stolyar S V., Bayukov O A, Gurevich Y L, Ladygina V P, Iskhakov R S and Pustoshilov P P 2007 Mössbauer study of bacterial ferrihydrite *Inorg. Mater.* **43** 638–41

[13] Raikher Y L and Stepanov V I 1992 Thermal fluctuation effect on the ferromagnetic-resonance line-shape in disperse ferromagnets *JETP* **102** 1409–23

[14] Poperechny I S and Raikher Y L 2016 Ferromagnetic resonance in uniaxial superparamagnetic

particles *Phys. Rev. B* **93** 014441
[15] Wajnberg E, El-Jaick L J, Linhares M P and Esquivel D M S 2001 Ferromagnetic Resonance of Horse Spleen Ferritin: Core Blocking and Surface Ordering Temperatures *J. Magn. Reson.* **153** 69–74

The effect of magnetic nanoparticle concentration on the structure organisation of a microferrogel

A V Ryzhkov[1,2], **P V Melenev**[2,3], **M Balasoiu**[4,5], **Yu L Raikher**[1,2]

[1] Perm National Research Polytechnic University, Perm, 614990, Russia
[2] Institute of Continuous Media Mechanics – the Division of Perm Federal Research Center, Russian Academy of Sciences, Ural Branch, Perm, 614013, Russia
[3] Ural Federal University, Ekaterinburg, 620083, Russia
[4] Laboratory of Neutron Physics, Joint Institute of Nuclear Research, Dubna, Moscow region 141980, Russia
[5] Horia Hulubei National Institute of Physics and Nuclear Engineering, Bucharest-Magurele, Romania

E-mail: `ryzhkov.a@icmm.ru`

Abstract. Coarse-grained molecular dynamics simulation is applied to study the structural response of micro-sized magnetopolymer objects – microferrogels (MFG). The results for MFGs with different magnetic properties and concentrations of magnetic filler nanoparticles are analysed to detect the transition between non-aggregated configurations and the states with pronounced chains. The nanoparticles are assumed to be either magnetically isotropic or to possess infinite magnetic anisotropy. It is shown that, depending on the type of the particle anisotropy, an applied field in rather different ways affects the MFG structure and shape. Diagrams describing the degree of aggregation as a function of the parameter of the interparticle magnetodipolar interaction and concentration are presented. In particular, it is found that in the case of infinitely anisotropic nanoparticles the aggregation transitions undergoes via a non-trivial scenario. The effect of the structure transformations on the volume change of the MFG objects is studied as well.

1. Introduction

Microferrogels (magnetic microgels, microscopic ferrogels, MFGs) are isolated soft polymers objects with typical sizes in the range of 0.1-10 μm containing embedded magnetic nanoparticles (NPs); typical NP materials are magnetite and cobalt ferrite [1, 2]. Strong coupling between the magnetic NPs and the polymer mesh, in which the NPs take on the role of cross-linkers, ensures the dependence of mechanical and structural properties of a MFG object on the external magnetic field. Therefore, an inherent property of MFGs is the remote control over their structure, volume and shape. Such a possibility of field-aided manipulations is very important, for example, for drug delivery, drug release and other biomedical applications [3, 4, 5, 6].

As it was shown in Refs. [7, 8], structure (i.e., chain formation), magnetisation and volume changes due to application of an external uniform magnetic field depend on the type of uniaxial magnetic anisotropy of NPs. However, the above-cited works were focused exclusively on highly-filled samples with strong dipolar magnetic interaction between NPs. Hereby we continue molecular dynamics study of magneto-structural and magneto-mechanical responses of individual MFGs and analyse the dependence of structural properties of MFGs on concentration of the magnetic filler particles as well as on the parameter of interparticle interaction.

2. Model

We use two models of nanoparticles which correspond to the limiting cases of their internal magnetic anisotropy. In the first case, it is a particle whose magnetic moment is entirely free to rotate inside its body, i.e., the particle has zero anisotropy and, thus, is perfectly magnetically soft. The other model implies that the magnetic moment is tightly "frozen" in the particle body, i.e., the particle is absolutely magnetically hard, its magnetic anisotropy is infinite. To distinguish between those two situations, we define parameter $\sigma = E_A/k_BT$ that relates the magnetic anisotropy energy to thermal one. Therefore, the considered particles have either $\sigma = 0$ or $\sigma = \infty$.

The model MFG samples generated for simulations contain the particles of the same σ, which make the nodes of a simple cubic lattice constructed of bead-spring polymer strands. In the framework of coarse-grained molecular dynamics, each strand is constructed of a chain of structure units (polymer blobs) which, but conditionally, are termed as "monomers". Each monomer has two neighbours, to which it is linked by central-bonded harmonic potentials equivalent to simple springs with equal rigidity and equilibrium lengths. In an MFG, the end monomers of some strands are attached in the same way to the NPs. Under such linking, the MFG subsystems turn out to be always coupled, while, as shown below, the parameter σ determines the extent of that coupling.

The concentration of magnetic filler in the samples is defined on the basis of the number of the lattice (mesh) nodes occupied by the particles. The residual fraction of the nodes are the sites where the polymer chains are connected only with each other; in simple cubic lattice this means a "meeting point" of six chains.

The samples with different fractions of magnetic filler were generated in the following way. First a "pre-initial" configuration, being a simple cubic lattice with 10^3 nodes, is used, where the lattice period is equal to the contour length of the polymer strand. In our simulations all the strands were equal and comprised 12 identical blobs (monomer). To make the polymer object sufficiently soft and to provide mobility for the magnetic particles, the size of the monomer is taken three times smaller that that of the NP., and no angular rigidity is introduced.

As soon as the magnetic phase concentration c for a given simulation is defined, then the numbers of the occupied and particle-free nodes are calculated as $N_p = cN_{tot}$, where N_{tot} is the total number of nodes in the sample. Thus determined amount of NPs is distributed at random over all the nodes.

After this "pre-initial" configuration of the sample is assembled, it is subjected to thermalisation by setting it in the thermostat of given temperature and letting the polymer strands to fluctuate. This thermalisation stage is also preliminary, as during it the NPs are treated as non-magnetic, and when performing simulation, only steric potentials for the structure units (blobs and NPs altogether) and the bonding potentials of the blob strands are taken into account. To allow for the stochastic nature of the thermostat action as well as for random distribution of NPs over the mesh nodes, the averaging over several implementations of the same sample is carried out. The snapshots of MFG samples with different particle concentrations are shown in Fig. 1.

Upon having attained the thermalised state, the magnetodipole interaction between the NPs, all of which are assumed to bear equal permanent magnetic moments μ and be of the same diameter d_{np}, is turned on. The intensity of this interaction is characterised by the nondimensional parameter $\lambda = \mu^2/(d_{np}^3 k_BT)$ that relates the magnetodipole energy of two particles in tight contact to thermal energy. Then the final stage of simulation is commenced: the equation of motion of all the elements of the MFG sample are integrated (with all the

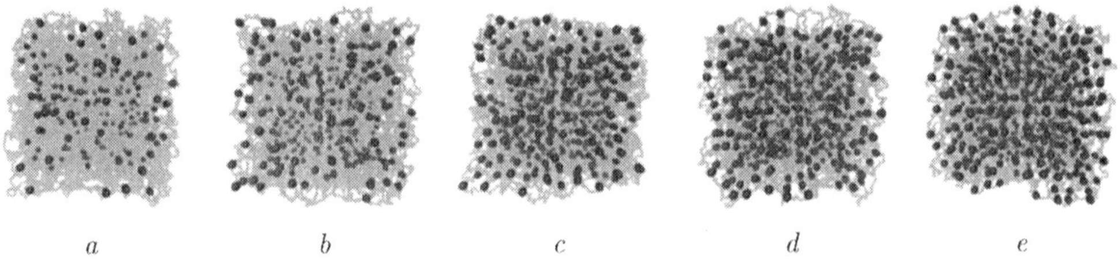

Figure 1. (Colour online) Thermalised configurations with different number of nodes occupied by NPs; 20% of nodes (a), 35% (b), 50% (c), 65% (d), 80% (e).

constraints and random forces and torques) until the energetically stationary state is attained.

3. Results and discussion

In this work we focus on the structures induced in the two afore-mentioned types of MFGs (with $\sigma = 0$ or ∞) in the absence of external magnetic field. For that, we use an indicator function $q(j, \tilde{d})$ introduced in Ref. [7]. It renders the fraction of NPs possessing at least j neighbours whose center-to-center distances are less or equal the value $\tilde{d}$ that is chosen quite close to d_{np}; in our calculations we set $\tilde{d} = 1.1d_{np}$. Therefore, the values $q\left(1, \tilde{d}\right)$, $q\left(2, \tilde{d}\right)$, and $q\left(3, \tilde{d}\right)$ yield the number of particles which have at least one, two or three nearest neighbours, respectively.

As a general fact, we remark that in all the MFGs studied, the value of $q\left(3, \tilde{d}\right)$ turns out to be virtual zero that points out the absence of multi-particle clusters and/or branched NP aggregates. In other words, practically all the forming aggregates are chains.

In all the simulations we find that for a given concentration, the value of $q\left(1, \tilde{d}\right)$ that reflects the presence of any aggregates (including pair clusters) grows with the increase of the dipolar interaction λ. As shown in Fig. 2, indicator $q\left(1, \tilde{d}\right)$ changes mostly fast (from 0.2 to 0.8) in a relatively narrow interval of λ. Because of that, one may interpret the occurring change of the MFG as a structure transition from a nearly homogeneous (the majority of the particles are single) to the aggregated state where the majority of the particles are united in chains. It seems reasonable to associate the finite width of this interval (at c =const) with the distributions of (i) the local interparticle distances and (ii) the local effective elasticity of the mesh in a given realisation of the MFG sample; fluctuations of those distributions in different realisations of the sample should also contribute to smearing of the transition region. We remark that in a given MFG the only way to initiate the above-discussed transition is to move parameter λ by varying temperature. However, temperature is a factor that influences all the properties of a gel that has to be taken into account for real systems.

From the diagrams of Fig. 2 it is seen that the transition zones are bow-shaped (leftward inflected) independently of the type of the particle anisotropy. Therefore, in any of MFGs under consideration the aggregation transition requires minimal intensity of the interparticle interaction at moderate concentrations of the magnetic phase, while to induce the same transition at either low or high concentration one needs to enhance the value of λ.

Note that the bow-shaped configuration of the transition zones in Fig. 2 is due to two different effects. At the low-concentration end it is entailed by the fact that mean interparticle distances in such MFGs are large. As the magnetodipole coupling decreases rather fast with the distance, to provide a sufficient attraction force at given temperature it is necessary to enhance the reference value of the interaction, i.e., to increase λ in fact is. In a system with high NP concentration, the factor impeding aggregation is most probably the self-averaging of the local field. Mutual

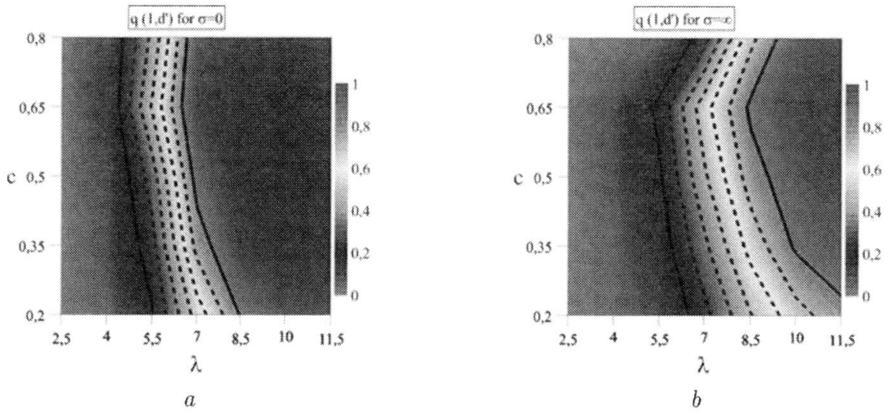

Figure 2. (Colour online) Dependence of the aggregation indicator $q\left(1,\tilde{d}\right)$ on concentration c and parameter of magnetodipolar interaction λ for: (a) magnetically isotropic particles $\sigma = 0$ and (b) magnetically anisotropic particles $\sigma = \infty$. Solid black lines are isolines of $q\left(1,\tilde{d}\right) = 0.2$ and 0.8, dashed black lines correspond to intermediate values of $q\left(1,\tilde{d}\right)$.

attraction of a given pair of particles is slackened by the fields induced by the surrounding ones which are positioned quite close to the pair. To overcome this screening, the enhanced pairwise interaction is required.

Comparison of the diagrams for MFGs with $\sigma = 0$ and $\sigma = \infty$ in Fig. 2 reveals that for anisotropic particles the aggregation is notably hampered. To understand the origin of this effect, it suffices to recall that in a system with $\sigma = \infty$ the orientation of the NP magnetic moment is strongly coupled to the positions of the blob chains grafted to the NP surface. Therefore, to point its magnetic moment along the direction optimising the magnetodipole interaction with other particles, the given one has to move its mesh environment. Evidently, this displacement requires some extra energy in comparison with the case of magnetically soft NP where the internal rotations of the magnetic moment do not cost any energy. This causes apparent difference between Figs. 2 a and b: the enhanced width of the transition "band" and a notable enhancement of the reference values of λ corresponding to the aggregation transition for any given concentration. On this basis, one concludes that, if to take two MFG samples possessing the same concentration and same magnitude of parameter of λ, then under zero external field the sample with anisotropic particles would display a considerably lower degree of aggregation.

The above-described character of the transition from non-aggregated to chained state in MFGs is confirmed by the plots of indicator $q\left(2,\tilde{d}\right)$, see Fig. 3. This function characterise the fraction of NPs which are incorporated in chains comprising at least three-particle clusters.

Aggregation of the particles resulting in formation of "coil" structures, see Fig. 3, affects the polymer conformations in the MFGs. By that, it entails the change of the sample density and, thus, its volume V. For MFGs with magnetically isotropic filler the volume change behaves expectably: the greater the magnetodipolar interaction, i.e., the higher the degree of chaining, the larger is the relative "volume defect" $\Delta V = (V - V_0)/V_0$ where V_0 is the volume of a thermalised sample with the switched off magnetic interactions. Under growth of the interparticles magnetic forces, i.e., of parameter λ, the volume defect ΔV of such a sample saturates due to pure sterical reasons. This is illustrated by Fig. 5a, see the dashed lines there.

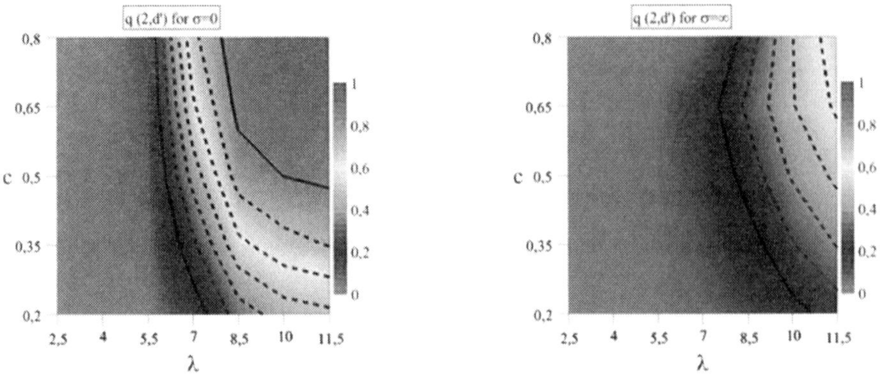

a *b*

Figure 3. (Color online) Dependence of the aggregation indicator $q\left(2, \tilde{d}\right)$ on concentration c and parameter of magnetodipolar interaction λ for: (a) magnetically isotropic particles $\sigma = 0$ and (b) magnetically hard particles $\sigma = \infty$. Solid black lines are isolines of $q\left(2, \tilde{d}\right) = 0.2$ and 0.8, dashed black lines correspond to intermediate values of the degree of aggregation.

Note that in this case the magnitude of ΔV increases monotonously with the NP concentration. We remark also that the behaviour reflected by Fig. 5a to a good extent correlates with the diagram of indicator $q\left(2, \tilde{d}\right)$ in Fig. 3a.

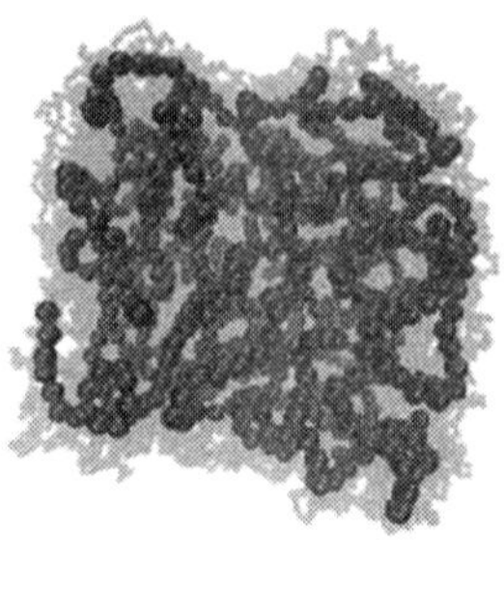

Figure 4. (Colour online)"Coil"-like aggregated configuration of an MFG that is densely filled ($c = 0.8$) with magnetically soft particles ($\sigma = 0$) possessing high dipolar moments ($\lambda = 11.5$)

For the model MFG with infinitely hard particles, the reference interval of the volume defect to the smaller value side, cf. colour scales in Figs. 5a and 5b. The cause of that is the necessity for the anisotropic particles to overcome rotational constraints imposed by the polymer strands in addition to the translational ones. Indeed, for the interaction parameters from $\lambda = 2.5$ to about $\lambda = 7.0$, the magnitude of ΔV grows rather slowly for low and moderate concentrations that implies that the distortion of the polymer in the vicinity of the "magnetic" nodes are yet small.

IOP Conf. Series: Journal of Physics: Conf. Series **994** (2018) 012004 doi:10.1088/1742-6596/994/1/012004

In the range of parameters where significant aggregation occurs ($\lambda > 7.0$) the volume defect changes faster but the saturation levels are attained at much larger values of the magnetodipolar parameter than in the case of an MFG with $\sigma = 0$.

Notably, at large NP concentrations for all the studied values of λ, the volume defect changes non-monotonically: its magnitude displays maximum about $c = 0.65$. In our view, for the weaker dipoles (say, $\lambda \leq 7$) such a feature is result of mutual competition of dipolar forces of closely positioned particles, i.e., the self-averaging of the local fields combined with lower compliance of anisotropic NPs to those forces. Under strong dipole interaction, the most probable reason of ΔV decrease is formation of rather extended NP chains which go through the whole sample thus making it less dense.

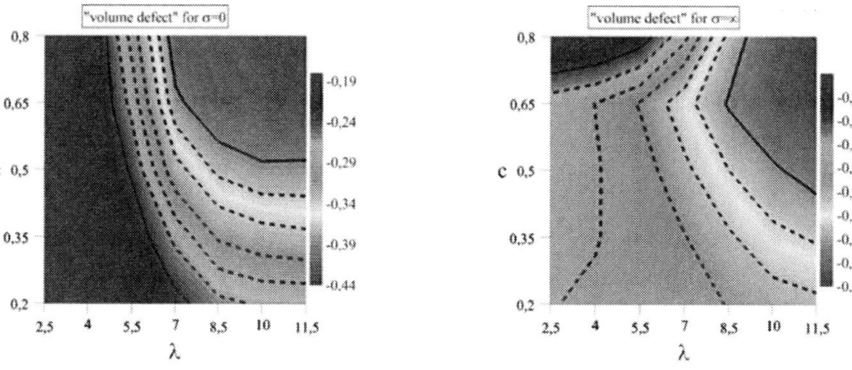

a *b*

Figure 5. (Colour online) Volume defect of an MFG sample as a function of concentration c and magnetodipolar parameter λ: (a) magnetically isotropic particles $\sigma = 0$ and (b) magnetically anisotropic particles $\sigma = \infty$. Solid black lines are isolines of volume, dot-dashed black line renders the average volume between the minimal and the maximal value at the diagram

4. Conclusions

Coarse-grained molecular dynamics simulations are performed of micro-size ferrogel objects with the reference number of magnetic nanoparticles being several hundreds. The structure state of the system might be non-aggregated as well as aggregated (chained) depending on the concentration of the magnetic phase and the intensity of interparticle interaction. With respect to the magnetodipolar parameter, the structure changes might be treated as a transition which is quite sharp for the microferrogels comprising magnetically soft particles and rather smeared for those comprising magnetically hard ones. The samples with the particles of different anisotropies possess substantially different magnitudes of the volume defect, i.e., effective density, as well. The results obtained show that the simulated MFGs can be used, in principle, to study the locking ability of microferrogels to carry microscale cargo since, under a fixed amount of incorporated particles, this ability depends on the structure of spontaneous configurations of the sample.

Acknowledgments

This work was supported by RFBR grants # 17-42-590504 and # 17-41-590160.

IOP Conf. Series: Journal of Physics: Conf. Series **994** (2018) 012004 doi:10.1088/1742-6596/994/1/012004

References

[1] Backes S, Witt M U, Roeben E, Kuhrts L, Aleed S, Schmidt A M and von Klitzing R 2015 *Journal of Physical Chemistry B* **119** 12129–12137
[2] Turcu R, Socoliuc V, Craciunescu I, Petran A, Paulus A, Franzreb M, Vasile E and Vekas L 2015 *Soft Matter* **11** 1008–1018
[3] Campbell S, Maitland D and Hoare T 2015 *ACS Macro Letters* **4** 312–316
[4] Li Y, Huang G, Zhang X, Li B, Chen Y, Lu T, Lu T J and Xu F 2013 *Advanced Functional Materials* **23** 660–672
[5] Thevenot J, Oliveira H, Sandre O and Lecommandoux S 2013 *Chemical Society Review* **42** 7099–7116
[6] Häring M, Schiller J, Mayr J, Grijalvo S, Eritja R and Daz D D 2015 *Gels* **1** 135–161
[7] Ryzhkov A V, Melenev P V, Balasoiu M and Raikher Yu L 2016 *Journal of Chemical Physics* **145** 074905
[8] Ryzhkov A and Raikher Yu L 2017 *Journal of Magnetism and Magnetic Materials* **431** 192–195

IOP Conf. Series: Journal of Physics: Conf. Series **994** (2018) 012005 doi:10.1088/1742-6596/994/1/012005

Neutron studies of paramagnetic fullerenols' assembly in aqueous solutions

V T Lebedev[1], A A Szhogina[1], M V Suyasova[1]

[1]B.P.Kostantinov Petersburg Nuclear Physics Institute, NRC Kurchatov Institute, 188300 Gatchina, Leningrad distr. Russia

levedev_vt@pnpi.nrcki.ru

Abstract. Recent results on structural studies of aqueous solutions of water-soluble derivatives of endofullerenes encapsulating 4f- and 3d-elements have been presented. Neutron small angle scattering experiments allowed recognize subtle features of fullerenols assembly as dependent on chemical nature (atomic number) of interior atom, pH-factor and temperature of solutions. It was observed a fractal-type fullerenols' ordering at the scale of correlation radii ~ 10-20 nm when molecules with iron atoms are integrated into branched structures at low concentrations ($C \leq 1$ % wt.) and organized into globular aggregates at higher amounts ($C > 1$ % wt.). On the other hand, for Lanthanides captured in carbon cages the supramolecular structures are mostly globular and have larger gyration radii ~ 30 nm. They demonstrated a good stability in acidic ($pH \sim 3$) and neutral ($pH \sim 7$) media that is important for forthcoming medical applications.

1. Introduction

Since the discovery of La-endofullerenes [1] there is a growing interest in their synthesis and investigation of physical and chemical properties assuming also various biomedical applications, e.g. as Magneto-Resonance Imaging agents or antitumor preparations [2,3]. The unique endofullerenes' feature which is very attractive for medicine is a durable encapsulation of heavy metal atoms inside a carbon cage which is strongly resistant against chemical attacks in living organisms. These integrated atoms do not release out of cages that prevent a contamination of biological tissues. At the same time, such isolated atoms retain their native properties, e.g. magnetic ones creating high relaxation rates of surrounding protons in biological medium where (MRI-diagnostics) [4]. On the other hand, a key problem of successful biomedical application of endometallofullerenes (EMF) is related to providing their solubility in aqueous media and this stimulates a search of ways to limit or regulate possible aggregation (clustering) to prevent irreversible sedimentation. The aim of this work is a comparative study of endofullerenols' solutions ordering when their molecules contain 4f- or 3d-elements.

2. Experimental

Two series of fullerenols have been used [5,6]. The first group of the samples has included iron based endofullerenols $Fe@C_{60}(OH)_X$ ($X \sim 30$) (Fig.1) prepared by the process of hydroxylation of $Fe@C_{60}$ endofullerenes synthesized in electric-arc in inert atmosphere (helium) by the evaporation of composite electrodes (mixture of graphite and iron phthalocyanine pyrolysate).

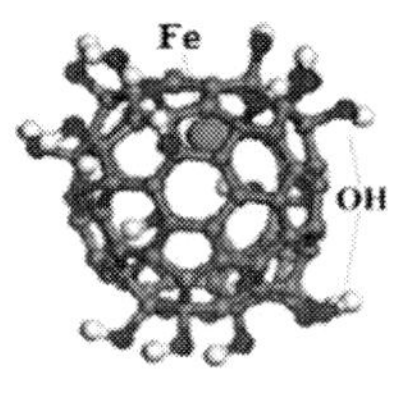

Figure 1. Schematic view of iron endofullerenol $Fe@C_{60}(OH)_X$.

In small angle neutron scattering experiments the crucial features of iron fullerenols self-assembly in aqueous solutions have been studied ("YuMO"-spectrometer, JINR, Dubna) for the concentrations C = 0.25-2.0 % wt. at ambient (20°C) and enhanced temperature (37°C, physiological conditions). In the experiments it was varied the pH value ($pH \sim 3$, fullerenols in pure water; $pH \sim 7$, phosphate buffer) to detect expected fullerenols' local ordering which can be regulated in this way.

The second group gathered fullerenols $M@C_{2n}(OH)_Y$ ($Y \sim 40$) with 4f-elements (M = Pr, Sm, Eu, Gd, Tb, Dy, Ho, Tm). These substances contained of 5-10 % wt. of metal, and their carbon cages C_{2n} were mainly C_{82}. The presence of higher fullerenes ($2n > 82$) was negligible since in the soot their amount was detected of 1.5 % wt. only.

In general it was used a common procedure of samples' preparation. All fullerene-based structures embedded in the soot produced in electric arc then were extracted by the solvent (dimethylformamide) with a special catalyst addition. The soluble substances served for a production of fullerenols in the reaction of hydroxylation by hydrogen peroxide. Finally the mixtures of fullerenols were purified from the solvent traces during a double process of recrystallisation.

The chemical structure of fullerenols has been tested by IR-spectroscopy which confirmed OH-groups' bonding to carbon cages. All the molecules of fullerenols have had the number of hydroxyls of ~ 30-40 that provided their good solubility in light water. The aqueous solutions possessed of $pH \sim 3.7$-4.2 (acidic media) weakly dependent on atomic number of captured metal atom. It means a cleavage of protons from the molecules which got a negative charge. Nevertheless, electrostatic repulsive forces did not prevent fullerenols' aggregation of in various forms observed by small angle neutron scattering (SANS-Membrane, NRC KI – PNPI). Additionally the Atomic Force Microscopy (set up Solverbio, NT-MDT, semi-contact regime, probe NSG03) has been applied to visualize immediately a formation of fullerenols' supramolecular structures. Very diluted solutions (C ~ 10^{-2} % wt.) were deposited on polycrystalline Si, then dried and rinsed.

3. Discussion

SANS-data for iron fullerenols' aqueous solutions qualitatively are similar at different conditions (C, pH, temperature). For illustration the spectra for these samples at ambient temperature and $pH \sim 7$ are shown only (Fig.2).

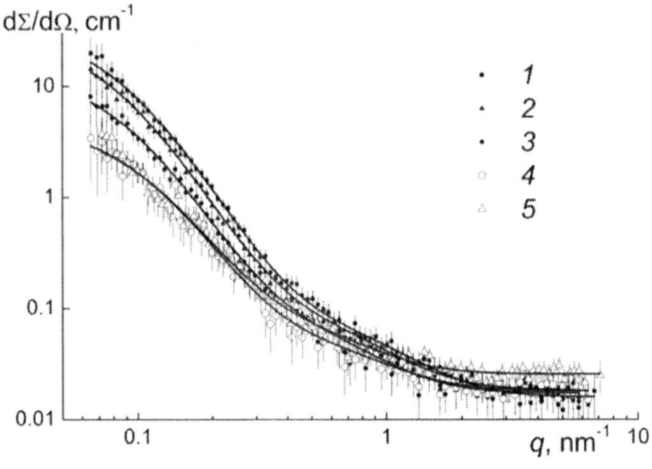

Figure 2. SANS-patterns for iron fullerenols aqueous solutions (20°C, $pH \sim 7$) for the concentrations C = 2.0, 1.5, 1.0, 0.5, 0.25 % wt. (1-5).

At low momentum transfers $q < 1$ nm^{-1} the samples demonstrate a high increase of cross sections, by three orders in magnitude. It indicates a strong clustering in diluted and more concentrated systems that is described by the scattering function

$$d\Sigma/d\Omega = I_o/[1 + (qR_C)^2]^{D/2} + Bg. \quad (1)$$

The fitting parameters I_o, R_C, D denote forward cross section, correlation radius and fractal dimension of observed structures, and incoherent background is presented by the parameter $Bg \approx$ const. The magnitude of $I_o \sim C$ increases proportionally to the concentration. It shows almost stable local ordering fullerenols in the concentration range C = 0.25-2.0 % wt. As it was found from the estimates of scattering ability of fullerenols, the aggregation numbers in solutions achieve the magnitudes of ~ 10^4 at spatial scale of $R_C \sim$ 10-20 nm (Fig.3). With the increase of concentration in neutral medium ($pH \sim 7$) it was observed a progressive growth of the radii $R_C(C) \sim$ 11-15 nm as a result of structural transformation when branched aggregates (fractal dimension $D \sim 2.5 < 3$) become of globular type ($3 < D < 4$) (Fig.3). These trends are revealed also in acidic medium ($pH \sim 3$) where the formation of aggregates is still more pronounced and they are greater, $R_C \sim$ 20 nm (Fig.4).

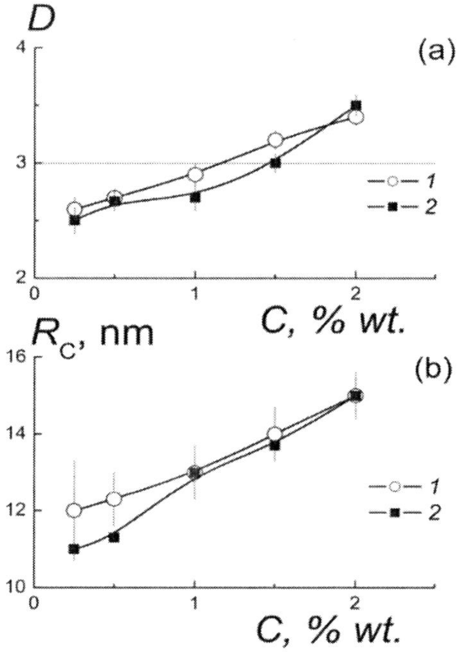

Figure 3. Fractal dimension $D(C)$ and correlation radius $R_C(C)$ (a,b) of fullerenols' structures vs. concentration in solutions ($pH \sim 7$) at 20°C and 37°C (1,2).

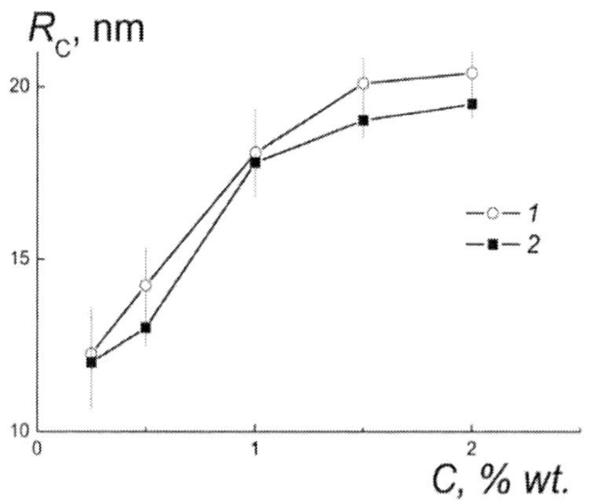

Figure 4. Correlation radius $R_C(C)$ of fullerenols' structures vs. content in solutions ($pH \sim 3$) at 20°C and 37°C (1,2).

A comparative analysis of scattering data for iron fullerenols in solutions at different conditions (neutral or acidic medium, 20°C and 37°C) has testified on relatively stable scale of molecular self-assembly by a transition from diluted to moderately concentrated systems when it was observed only a densification of branched structures which became globular entities.

In following experiments another specific features in the local ordering of fullerenols due to the presence of 4f-elements (Pr, Sm, Eu, Gd, Tb, Dy, Ho, Tm) revealed at fixed concentration in solution (C = 1.5 % wt.). The data allowed understand in which way a variation of atomic numbers of metals (Z = 59-69) has modified fullerenols' assembly. For fullerenols with Lanthanides the SANS-data transformed into the correlation functions $\gamma(R)$ exhibited qualitatively similar patterns of $P(R) = R^2\gamma(R)$ in the range of radii R = 0-90 nm (Fig.5).

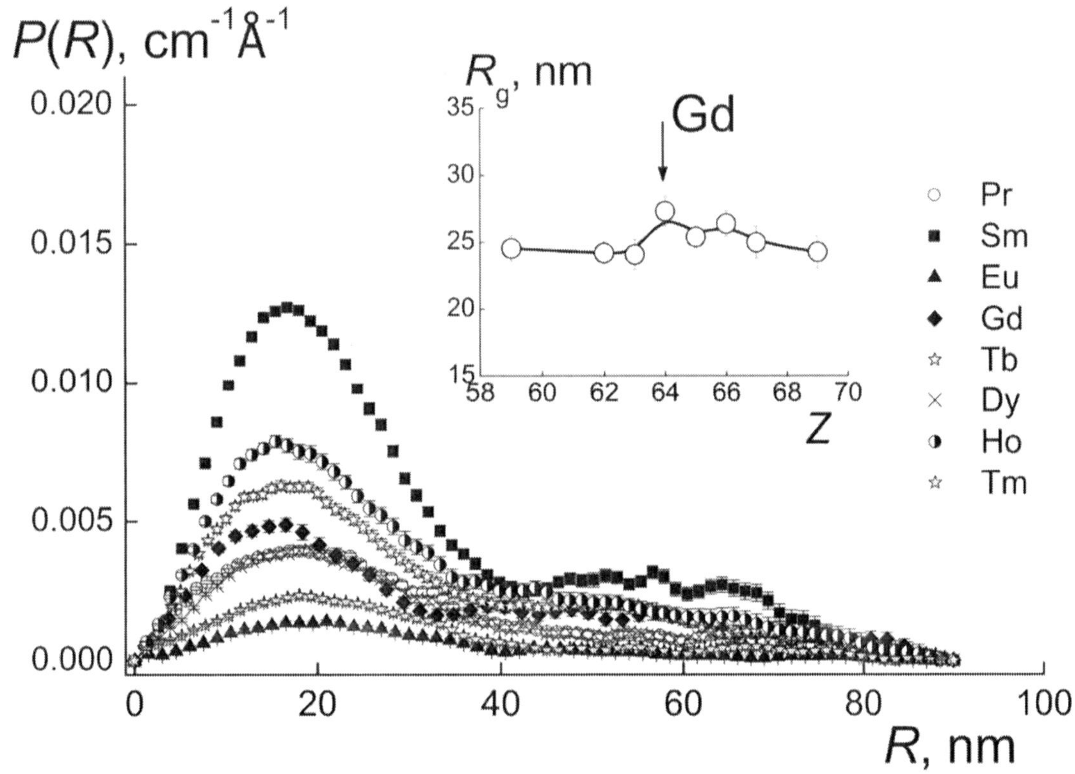

Figure 5. Fullerenols $M@C_{82}(OH)_X$ (X ~ 38-40) with 4f-metals (Pr, Sm, Eu, Gd, Tb, Dy, Ho, Tm). Spectra $P(R)$ for aqueous solutions: concentration 1.5 % wt., ambient temperature (20°C), pH = 3.

A typical set of spectra $P(R)$ is presented for fullerenols in pure water (pH ~ 3) at ambient temperature (20°C). The increase of temperature up to 37°C and transfer to neutral conditions (pH ~ 7) did not cause any substantial structural changes in fullerenols' assembly. All the spectra for fullerenols captured the elements in the diapason of atomic numbers Z = 59-69 have exhibited qualitatively similar maxima in the spatial interval of R = 0-90 nm. The most intense peak located at R_m ~ 15-20 nm indicates a characteristic radii of aggregates.

The row of rare earth fullerenols demonstrates a common scale of molecular order in aqueous solutions. Fine structural features of the ordering of molecules with different metals inside are imprinted in the behavior of average gyration radii $R_g(Z)$ (Fig.5, inset). It gets a maximum for Gd-fullerenols which possess the most pronounced paramagnetic properties of fundamental and practical interest for MRI (Fig.5). On the other hand, in the row of Lanthanides it was observed a substantial dispersion in aggregation degrees ~ 10^3-10^4 those are in the amplitudes of spectral maxima. The lowest ability to form nanostructures was detected for Eu-fullerenols, and the highest one was found for Samarium-based endohedral molecules, while in the case of captured Gadolinium it was observed a moderate aggregation (Fig.5).

IOP Conf. Series: Journal of Physics: Conf. Series **994** (2018) 012005 doi:10.1088/1742-6596/994/1/012005

4. Summary

A comprehensive SANS-analysis of the forms of molecular ordering for different kinds of fullerenols encapsulated 3d- or 4f-elements has elucidated really common features of molecular assembly observed as a formation of branched or globular superstructures relatively stable by a variation of *pH*-factor at ambient or physiological temperature. This demonstrates a universal character of fullerenols' interactions in aqueous solutions which structuring is governed mainly by hydrogen bonds between hydroxyl-shells around carbon cages while a variation of chemical nature of metal inside a cage induces relatively weak effects in fullerenols' interaction, e.g. due to electric or magnetic dipolar forces like in the case of Gd-fullerenols which showed a maximum gyration radius of aggregates among the structures with different Lanthanides.

Acknowledgement
Authors are grateful to Dr. A.I. Kuklin for technical assistance.

References

[1] Heath J R, O'Brien S C, Zhang Q, Liu Y, Curl R F, Tittel F K, Smalley R E 1985 *J. Am. Chem. Soc.* **107** 7779

[2] Piotrovsky L, Eropkin M, Eropkina E et al. / 2008 eds F Cataldo, T Da Ros // Medicinal Chemistry and Pharmacological Potential of Fullerenes and Carbon Nanotubes (Germany: Springer, p 139–155)

[3] Krokosz A, Grebowski J, Rodacka A, Pasternak B, Puchala M 2014 *Radiation Physics and Chemistry* **97** 102

[4] Koltover V K 2008 *Vestnik RFFI* **59(3)** 54 (Rus.)

[5] Lebedev V T, Szhogina A A, Bairamukov V Yu 2017 *IOP Conf. Series: Journal of Physics: Conf. Series* **848** 012005 doi :10.1088/1742-6596/848/1/012005

[6] Lebedev V T Kulvelis YuV, Runov V V, Szhogina A A, Suyasova M V 2016 *Nanosystems: physics, chemistry, mathematics* **7(1)** 22

Studies of electrochemical interfaces by TOF neutron reflectometry at the IBR-2 reactor

V I Petrenko[1,2*], I V Gapon[1,2], A A Rulev[3], E E Ushakova[3,4], E Yu Kataev[3], L V Yashina[3], D M Itkis[3,5], M V Avdeev[1,5]

[1]Frank Laboratory of Neutron Physics, Joint Institute for Nuclear Research, Dubna, Moscow Reg., Russia
[2]Taras Shevchenko National University of Kyiv, Kyiv, Ukraine
[3]Department of Chemistry, Lomonosov Moscow State University, Moscow, Russia
[4]Engineering Incubator Ltd., Dubna, Moscow Reg., Russia
[5]Dubna University, Dubna, Moscow Reg., Russia

E-mail: vip@nf.jinr.ru

Abstract. The operation performance of electrochemical energy conversion and storage systems such as supercapacitors and batteries depends on the processes occurring at the electrochemical interfaces, where charge separation and chemical reactions occur. Here, we report about the tests of the neutron reflectometry cells specially designed for operando studies of structural changes at the electrochemical interfaces between solid electrodes and liquid electrolytes. The cells are compatible with anhydrous electrolytes with organic solvents, which are employed today in all lithium ion batteries and most supercapacitors. The sensitivity of neutron reflectometry applied at the time-of-flight (TOF) reflectometer at the pulsed reactor IBR-2 is discussed regarding the effect of solid electrolyte interphase (SEI) formation on metal electrode surface.

Keywords: electrochemical interfaces; neutron reflectometry; specular reflectivity; lithium ion batteries; liquid electrolytes; solid-electrolyte interphase.

1. Introduction

Electrochemical energy conversion and storage systems often play a crucial role in overall performance of a wide variety of devices that we face in our everyday life ranging from portable phones, laptops and power tools, to rather big facilities including power / heat cogeneration fuel cell systems, railway locomotives, power grid leveling facilities. The operation of the most of these systems such as supercapacitors [1], rechargeable batteries [2] is based on the electrochemical interfaces, where charge separation and chemical reactions occur. The evolution of the structure, composition and chemistry at the interfaces between electrodes and electrolytes affects all the functional parameters of the device including power and long-term performance stability. That motivates the researchers to study the chemical processes at the interface itself and transport phenomena in bordering layers under operando conditions. The methods that allow effective monitoring of internal electrode-electrolyte interfaces in operando conditions are of current interest [3]. Among such methods, neutron reflectometry (NR) is one of few techniques that provide a sensitivity to such interfaces due to high penetration power of thermal neutrons into matter. An additional important feature of NR is a possibility to vary 'electrode-

IOP Conf. Series: Journal of Physics: Conf. Series **994** (2018) 012006 doi:10.1088/1742-6596/994/1/012006

electrolyte' scattering contrast based on isotope substitution of hydrogen for deuterium in the liquid component to better resolve near-surface layers of interest.

An in-deep understanding of atomistic mechanisms underlying different electrochemical process, including electric double layer formation, Faradic processes involving heterogeneous electron or ion transfer, often require that data are to be recorded *in situ*. *Ex situ* studies in lots of cases are almost irrelevant, since the chemical reactions occurring during operation of electrochemical devices result in intermediates and products that cannot be quenched for post-process studies. The desired parameters include (i) the chemical composition at the interface, (ii) the chemical state of atoms and it is variation due to the chemical / electrochemical reactions, and (iii) structural evolution in the interface region.

Here, we report about the tests of the NR cells specially designed for operando studies of the structural changes at the electrochemical interfaces between solid electrodes and liquid electrolytes. The cells are compatible with anhydrous electrolytes with organic solvents, which are employed today in all lithium ion batteries and most supercapacitors. The concept of the cells mainly repeats the ideology of the previous devices for this purpose [4, 5]. It uses the horizontal layout of the interface with the neutron beam passing through a massive single crystal silicon block and falling to a thin (thickness below 100 nm) electrode film on silicon which is in contact with a liquid electrolyte containing lithium. The experiment can be done when the electrodes are in close electric circuit under potentiostatic control. As a first step, homogeneous metal film (copper) is used as a working electrode The task of the experiment was to determine the possibilities of these units applied at the GRAINS time-of-flight (TOF) reflectometer [6], with horizontal sample plane at the IBR-2 pulsed reactor.

2. Experimental

The tested NR cells used the horizontal 'from the top' layout (Fig.1) or 'from the bottom' layout with the neutron beam passing through a massive (dimensions 8 × 5 × 1.5 cm) single crystal silicon block and falling to the interface between a thin electrode film (working electrode) on the silicon substrate and liquid electrolyte. As a working electrode, a metal (copper) film was used. The electrode deposition (magnetron sputtering) on several Si substrates was made at PNPI NRC KI (Gatchina, Russia) in two steps including the deposition of an intermediate Ti layer (nominal thickness 5 nm) for better copper adhesion and the deposition of the Cu layer itself (nominal thickness 50 nm) on the working silicon surface (plane orientation <111>, nominal roughness < 0.5 nm).

Several designs of NR cells have been developed. The main common features of the cells are demonstrated in Fig.2, where a cell for the from the top experimental configuration is presented. The cell is designed airtight with o-ring sealing between silicon single crystal and the cell enclosure. The sealing is important as the electrolytes used should contain no more than 10 ppm humidity. The cell enclosure is fabricated of polyether ether ketone (PEEK), the polymer which comprises proper mechanical properties together with high chemical stability in contact with organic solvents and does not react with metallic lithium that is used as a counter electrode (Li foil from China Energy Lithium, ltd., pressed on Ni mesh). The important feature of the cell is the introduction of Vycor-frit separate reference electrode (RE) that allows accurate potential control of the working electrode (WE). We use Ag/Ag^+ RE prepared in-house by sealing a silver wire in a tube with a solution of 0.01M $AgNO_3$ (Aldrich) and 0.1M tetrabutylammonium perchlorate (Aldrich, for electrochemical analysis) in acetonitrile (Aldrich, anhydrous). The RE potential was calibrated against a ferrocene / fferrocenium redox couple prior to the cell assembly. The cell was assembled including the electrolyte filling in the controlled atmosphere Ar-filled glovebox (humidity and oxygen concentrations below 0.1 and 1.0 ppm, respectively). The potential between the electrodes of the NR cell was controlled using a Bio-logic SAS SP-200 potentio / galvanostat in three-electrode configuration.

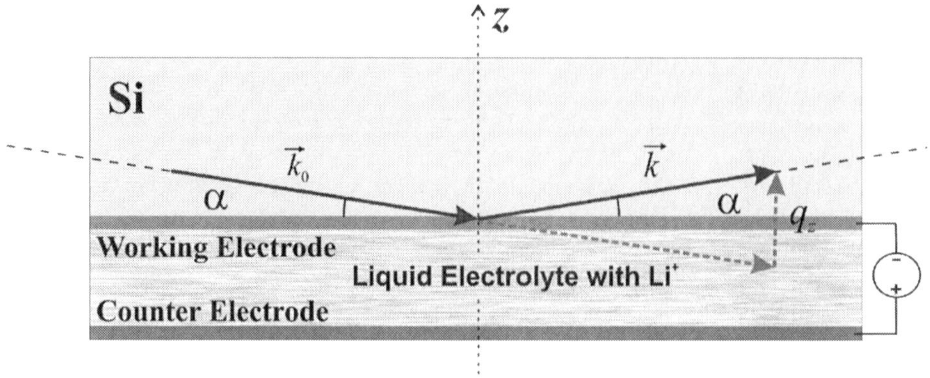

Figure 1. Principal 'from the top' layout of the NR experiment with the electrochemical cell. The plane neutron beam passes single-crystal Si and falls on interface between thin electrode film (working electrode) on the silicon substrate and liquid electrolyte. The ratio between the specular reflected (wave vector $\vec{k}$) and incident (wave vector $\vec{k_0}$) neutron beams known as reflectivity is measured as a function $R(q_z)$ of the Z-projection of the momentum transfer, $\vec{q} = \vec{k} - \vec{k_0}$, $q_z = (4\pi/\lambda)\sin\alpha \approx 4\pi\alpha/\lambda$, where λ is the neutron wavelength and α is the grazing angle. The potential between the working electrode (copper film) and the counter electrode (lithium foil) is remotely controlled.

In the first tests of the cells, 0.1 M solution of $LiClO_4$ (Aldrich, battery grade) in propylene carbonate (PC) was used as an electrolyte. Two kinds of solvents in the electrolyte were probed including protonated PC (h-PC,$CH_3C_2H_3O_2CO$) and deuterated PC (d-PC, $CD_3C_2D_3O_2CO$) both purchased from EQ Laboratories GmbH. The required electrolyte volume was about 6 ml (cell with the from the top configuration) and $<$ 3 ml (cell with the from the bottom configuration) per one cell filling.

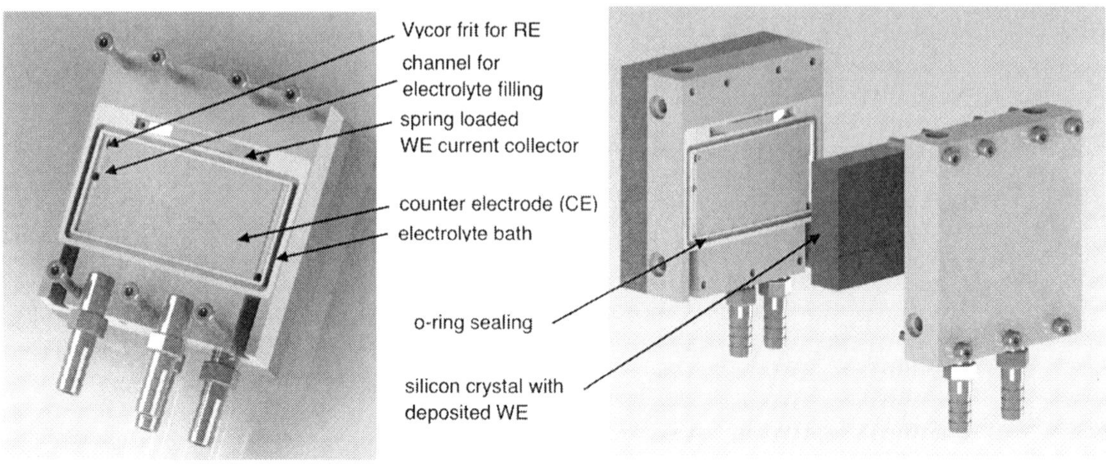

Figure 2. The design of electrochemical cell for neutron reflectometry, 'from the top' configuration.

The NR measurements were carried out at the GRAINS time-of-flight reflectometer at the IBR-2 pulsed reactor of FLNP JINR (Dubna, Russia) [7] in the thermal mode of the moderator.

The inclination ($\approx$ 5 mrad) of the plane-incident beam (collimated by several slit apertures) against the interface was provided by a deflecting neutron mirror (mirror-sample distance 3 m). The specular beam was detected by a 2D position-sensitive detector (size 20 $\times$ 20 cm^2, resolution 2 $\times$ 2 mm^2, sample-detector distance 3 m) [8] with the averaging of the reflected beam along the horizontal direction. The intensity was plotted as a function of the momentum transfer, $q_z = (4\pi/\lambda)\sin\theta \approx 4\pi\theta/\lambda$, where θ is the fixed (5 mrad) incident grazing angle (see Fig.2) and λ is the neutron wavelength varied in the working interval of 0.1 - 0.6 nm. The calibration against the incident spectrum was performed by a separate measurement of an empty beam (after removing the cell from the beam). The obtained NR curves were processed in terms of the Parratt formalism using the Motofit package for the IGOR Pro software [9].

3. Results

Prior the experiment with liquid electrolytes the structure of the pristine working electrode film deposited on the silicon substrate was determined by NR in the 'beam from air' experiments (Fig.3). Found from the fitting, the SLD profile (see inset to Fig.3) gives the best parameters collected in Table 1. The fitting procedure requires the introduction of an additional small transitional layer with respect to the boundary between Cu and Ti layers with high roughness and intermediate (between Cu and Ti) SLD-value. The parameters of the SLD profile of the working electrode film were fixed in the fitting of the next experimental curves obtained when the electrode was in contact with liquid electrolytes.

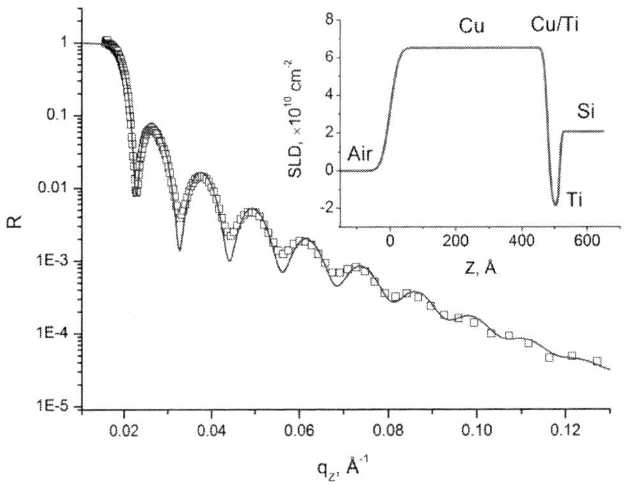

Figure 3. Experimental (points) neutron reflectivity curve for the silicon single crystal block with a deposited working electrode and the corresponding SLD profile (in the inset) derived from the experimental data fit (line); best parameters are collected in Table 1.

The NR experiment on the observation of the SEI layer for the cell filled with the electrolyte consists of two stages. First, the specular reflectivity curves are obtained under open circuit conditions. At the next stage, the deposition of the SEI layer on the copper electrode surface is initiated by applying to the cell a constant potential (here, + 10 mV vs. Li^+/Li during 15 min period). Then, the NR curves are obtained under open circuit conditions. The experimental reflectivity curves for the two (h- and d-) cases are collected and treated in Fig.4. Slight changes in the modulation of $R(q_z)$ because of the new small layer can be seen. The fits were made by varying reduced number of parameters of the interface layers including the thickness, roughness and SLD of the new layer, as well as the SLD of electrolyte. The last parameter was made free to be sure that the electrolyte SLD was not modified because of some parasitic effects (e.g. residual gas fraction) after cell assembly and filling.

Table 1. . Parameters of the SLD profile of the studied electrode on silicon substrate obtained by NR in the beam from air experiment.

Layer	Mean SLD, $\times 10^{10}$ cm^{-2}	Thickness, Å	Roughness, Å
Cu	6.5243	470	20
Cu/Ti	4.0	7	10
Ti	-1.9249	38	10
Si	2.0737	-	5

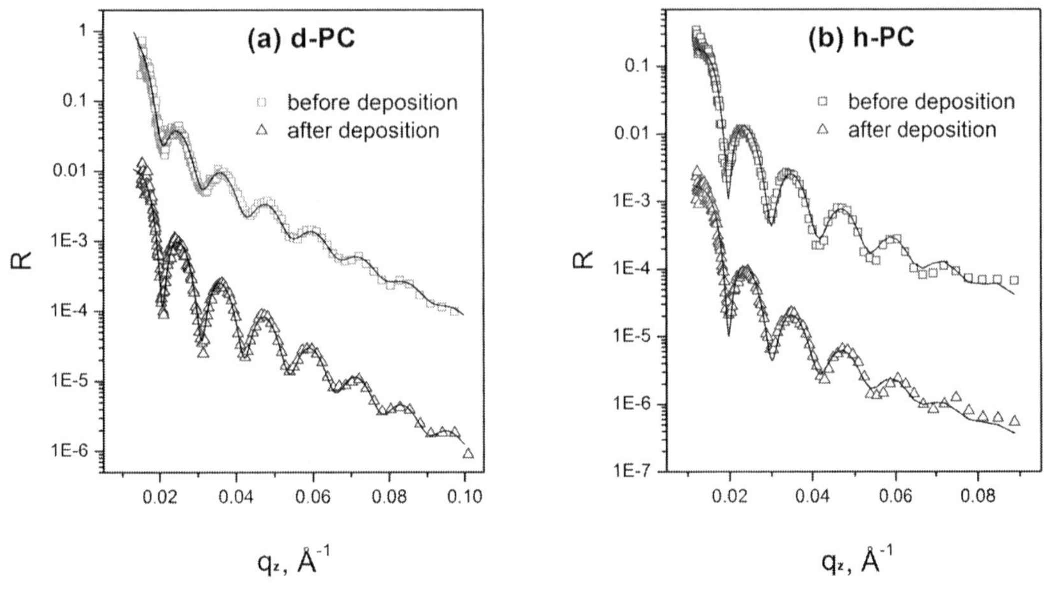

Figure 4. Experimental specular NR curves from electrochemical interfaces under study with d-electrolyte (a) and h-electrolyte (b) and fitting curves. The curves 'before deposition' and 'after deposition ' are shifted along the ordinate axis for convenient representation. To compare in absolute scale, the model curves are given Fig.5.

The effect of the SEI layer in the NR experiments with two kinds of electrolytes is analyzed in Fig.5, where the SLD profiles and corresponding model reflectivity curves obtained from the experimental NR curves before and after SEI deposition are compared. A new transitional (electrode-electrolyte) layer (characteristic thickness $\approx$ 50 Å) can be seen reflecting the SEI formation in agreement with the previous results for the similar interface [5]. As compared to the electrolyte, a lower SLD of this layer is a result of lithium enrichment; this decreases SLD since lithium has negative scattering length. While a modification of the SLD profile after SEI deposition takes place in both kinds of electrolyte, the effect is more pronounced in the case of the deuterated electrolyte. This is consistent with the scattering contrasts between electrolyte and electrode; the SLD difference is significantly less in the case of h-PC. The mean SLD of the SEI layers in the two electrolytes differs in absolute values showing the dependence of this parameter on the electrolyte SLD. The reason is that in the h-case hydrogen containing components of SEI decreases SLD of the layer because of low scattering length of hydrogen, while in the d-case the situation is opposite; SLD should increase because of the presence in it of deuterium with high scattering length. The additional potential factor determining this effect is a possible penetration of the liquid electrolyte into a porous at some extent structure of the SEI layer, thus increasing the effective content of hydrogen or deuterium in the layer for h- and d-solvents, respectively. To separate the contribution from the solvent, the so-called contrast variation should be done when experiments are repeated for the electrolytes with different deuteration rates and, hence, different SLDs and contrasts against the SEI layer.

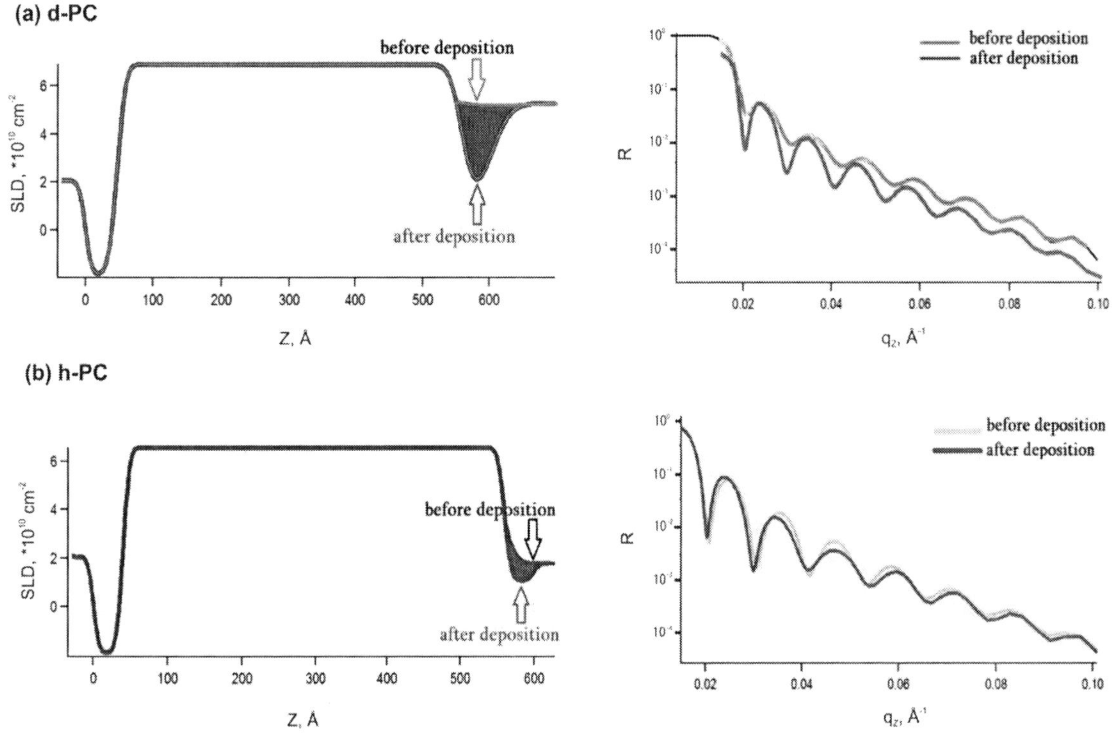

Figure 5. Effect of SEI layer in NR experiments with d-electrolytes (a) and h-electrolytes (b). The SLD profiles and corresponding model reflectivity curves obtained from the experimental NR curves before and after SEI deposition are compared. The differences in the SLD profiles before and after deposition are marked by red filling to illustrate different levels of deviations.

One can conclude that the q_z-range covered in the TOF experiments for the thermal (T = 300 K) mode of the neutron source carried out at one grazing angle is sufficient to reliably conclude about the appearance of a new layer (thickness above 40 $\mathring{A}$). It is important that this range is insensitive to the details of the inner titanium layer between the substrate and the electrode film followed from the XRR analysis. Our tests show that the SEI evolution on the metal electrode surface including thickness, roughness and mean scattering length density can be well followed in the operando experiments with deuterated electrolytes. In this connection, the 'from bottom' configuration of the NR cell is more preferable because of the lower volume of comparatively expensive deuterated solvents required in the experiment.

4. Conclusions

The neutron reflectometry experiments on electrochemical interfaces have been carried out using specialized cells for operando studies at the time-of-flight reflectometer. The interfaces between metal (copper) electrode and liquid electrolyte with lithium ions (lithium perchlorate dissolved in propylene carbonate) have been probed to clarify the sensitivity of the NR method to the formation of solid electrolyte interphase (SEI) on the metal electrode when analyzing specular reflectivity curves. The q_z-range covered in the experiments for the thermal mode of the neutron source (T = 300 K) is sufficient to reliably conclude about the appearance of a new layer (thickness above 40 $\AA$). The effect for light and heavy solvents in the electrolyte with respect to the D/H substitution have been compared in the same operando conditions. It is more pronounced for the heavy electrolyte because of the higher scattering contrast between electrode and electrolyte. The sensitivity of the scattering length density of SEI to the change in the electrolyte density after H/D substitution has been observed which is indicative of the electrolyte penetration into the structure of SEI.

Acknowledgments

The work was financed via Russian scientific foundation (project N 17-12-01540).

References

[1] Simon P, Simon P, Gogotsi Y and Gogotsi Y 2008 *Nat. Mater.* **7** 84585
[2] Tarascon J M and Armand M 2008 *Nature* **451** 652657
[3] Itkis D M, Velasco-Velez J J, Knop-Gericke A, Vyalikh A, Avdeev M V and Yashina L V 2015 *ChemElectroChem* **2** 1427–1445
[4] Lauw Y, Rodopoulos T, Gross M, Nelson A, Gardner R and Horne M D 2010 *Rev. Sci. Instrum.* **81** 074101
[5] Owejan E, Owejan J P, DeCaluwe S C and Dura J A 2012 *Rev. Sci. Instrum.* **24** 21332140
[6] Avdeev M V, Rulev A A, Bodnarchuk V I, Ushakova E E, Petrenko V I, Gapon I V, Tomchuk O V, Matveev V A, Pleshanov N K, Kataev E Y, Yashina L V and Itkis D M 2017 *Appl. Surf. Sci.* **424** 378–382
[7] Avdeev M V, Bodnarchuk V I, Petrenko V I, Gapon I V, Tomchuk O V, Nagorny A V, Ulyanov V A, Bulavin L A and Aksenov V L 2017 *Cryst. Rep.* **62** 1007–1013
[8] Belushkin A V, Bogdzel A A, Zhuravlev V V, Kutuzov S A, Levchanovski F V, Litvinenko E I, Jai L Y, Panteleev T T, Prikhodko V I, Chernikov A N, Churakov A V and Shevtsov V N 2010 *Phys. Solid State* **52**(5) 1025
[9] Nelson A 2006 *J. Appl. Cryst.* **39** 273–276

IOP Conf. Series: Journal of Physics: Conf. Series **994** (2018) 012007 doi:10.1088/1742-6596/994/1/012007

Exchange interactions and magnetic properties of hexagonal rare-earth-cobalt compounds

E.Burzo

Romanian Academy, Cluj-Napoca Branch, Romania

emil.burzo@phys.ubbcluj.ro

Abstract. The magnetic properties of some $Gd_xY_{1-x}Co_4A$ compounds with A = Co, Si or B are analysed including the pressure effects. Isomorphous structure transitions, parallelly with changes of cobalt moments from high spin states to low spin states, were shown as pressure increases. The magnetic data, obtained from band structures, were compared with those predicted by the mean field model.

1. Introduction

The magnetic properties of RCo_5–based compounds, where R is a rare earth or yttrium, were intensively investigated both from scientific reasons as well as in connection with their technical uses, as permanent magnets [1, 2]. These compounds crystallize in a $CaCu_5$-type structure, space group P6/mmm. In this lattice, the R atoms are located in 1a site and cobalt at 2c and 3g positions [1]. The RCo_4B compounds have a hexagonal structure, derived from $CaCu_5$ one, by replacing every two layers Co2c atoms by boron. There are two sites for R(1a, 1b) and Co(2c,6i) atoms and one for boron (2d) [3]. When R is a magnetic heavy rare-earth the above compounds are ferrimagnetically ordered, while for light rare earths, a parallel alignment of R and Co moments is shown. The easy axis of magnetization, for RCo_5 compounds with R = Y or Gd, is the **c**-axis, as imposed by the local uniaxial anisotropy of the 2c sites, higher than the planar one favoured by 3g atoms [4], YCo_4B shows also uniaxial anisotropy, at ambient temperature, which changes to planar one, at T_{SR} = 159 K [5].

The exchange interactions between magnetic rare-earths and cobalt are mediated by R5d band polarization, as suggested by the 4f-5d-3d model [6, 7]. The above polarization is the result of both local 4f-5d exchange as well as due to R5d-Co3d short range interactions, hybridization effects, respectively. The exchange interactions between cobalt atoms are of short range as well as those between R5d-R5d orbitals, in addition to those by means of conduction electrons (RKKY-type) [8].

In this paper, we analyse the exchange interactions in $Gd_xY_{1-x}Co_4A$ series with A = Co, Si, B in the framework of mean field model [9], as well as based on the data obtained from their band structures. The pressure dependences of the cobalt moments, for end series compounds, are theoretically investigated. A linear dependence of cobalt moments on the exchange splitting energy of their 3d bands was shown.

2. Experimental and computing method

The samples were prepared from high purity elements, in an induction furnace or by arc melting. An excess of rare earth (yttrium) of (2-4)% was added in order to compensate their losses during the melting. The samples were annealed at temperatures 900 °C ≤ T ≤ 1000 °C, in vacuum for one week. The X-rays analysis, at ambient temperature, evidenced the presence of only one phase. Some samples were already prepared [10, 11]. Magnetic measurements were performed in the temperature range 4.2 – 1000 K, and fields up to 70 kOe.

The ground state electronic structures and magnetic properties have been determined in the scalar-relativistic limit, i.e. without spin-orbit coupling, by using the tight-binding linear muffin-tin orbital (TB-LMTO) method within the atomic sphere approximation (ASA) [12, 13]. The local density approximation (LSDA) has been employed for the exchange and correlation potential of the electron gas, assuming Von Barth and Hedin parameterization [14]. The valence basis consists of s-, p-, and d-electrons. All *k*-space integrations were carried out with a 16×16×16×16 *k*-point mesh with an accuracy of 10^{-2} meV. The silicon was located in 3g sites, as previously reported [15].

3. Results

The partial density of states (DOS) of Co2c and Co6i atoms in YCo_4B, at ambient pressure, $v/v_0 = 1$, as well as at $v/v_0 = 0.88$ are plotted in Fig.1. The Fermi level, E_F, when $v/v_0=1$ is situated in a peak, or nearly on a peak in the density of states in spin-up cobalt sub bands at 2c and 6i sites, respectively. In the spin down sub-bands, the E_F is located in a deep in DOS. As effect of pressure, relative volumes reduction, respectively, there is a shift of the spin-up sub-bands to lower energy and of the spin down ones, at higher energy. As a result, there is a decrease of DOS at E_F in spin up sub band and an increase in spin down one. When cobalt in nonmagnetic, the densities of states for spin up and spin down sub bands are near identical, in a mirror-plane. The evolution with pressure of the centres of gravity, on the energy scale, for cobalt sub-bands were determined, the exchange splitting energy of Co3d bands, ΔE_s, respectively. A linear dependence of cobalt moments, at both sites, on the exchange splitting energy was shown, with a rate $\Delta E_s/M_{Co}=0.94$ eV/μ_B, close to the previous reported values [10, 16, 17]

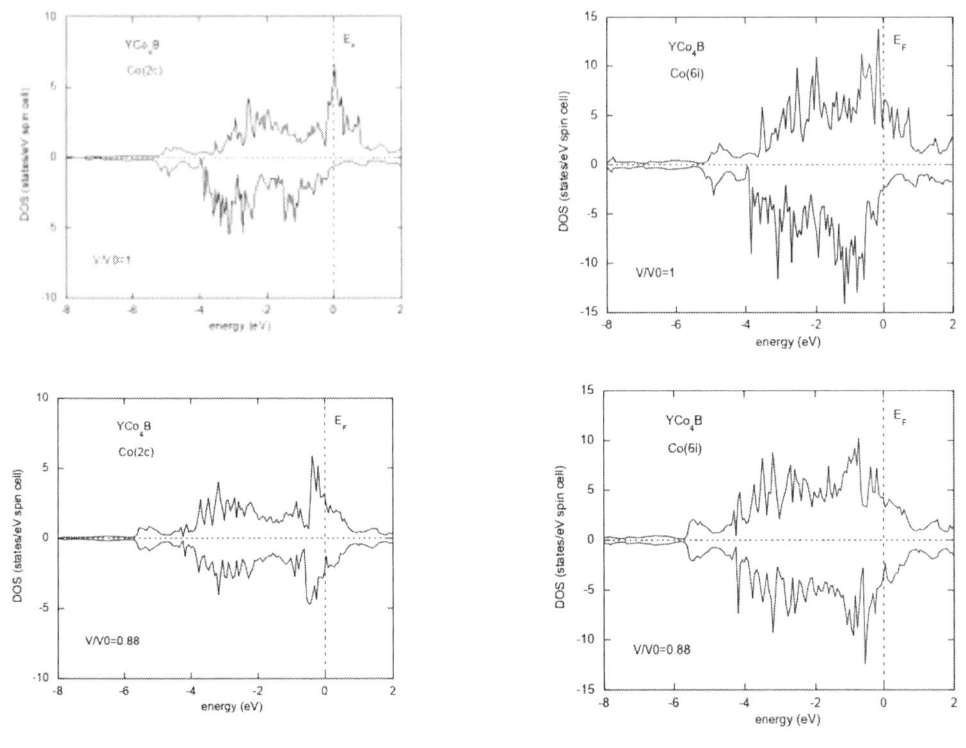

Figure 1. Partial densities of states for Co2c and Co6i atoms in YCo_4B compound, at $v/v_0=1.0$ and 0.88.

The pressure, relative volume dependences, respectively of the cobalt moments at 2c and 3g/6i sites in $GdCo_5$, YCo_4Si and YCo_4B are given in Fig.2. At relative low pressures, corresponding to reduced volumes 0.965 δ v/v_0 δ 1.0, there is a nearly linear decrease of cobalt moments, the corresponding rates ΔM/M being higher, as the initial cobalt moments are smaller. This behaviour can be correlated with their itinerancy degree – Fig.3. At a characteristic pressure, relative volume, respectively, a collapse of the cobalt moments from high spin state to low spin one is evidenced. When the initial cobalt moments are small, as example at 6i site in YCo_4B, a direct transition to nonmagnetic state is evidenced. The magnetic anomaly develops parallelly with a sudden decrease of a c/a ratio, an isomorphous phase transition, respectively. Similar behaviour has been already evidenced in YCo_5 compound and attributed [18] to a first order Lifshitz transition [19] or to a spin reorientation transition [20]. The transition develops, parallelly with the change of the Fermi surface topology. A change of anisotropy from uniaxial to planar, in YCo_5, has been previously suggested in connection with the diminution of cobalt moments [21-25]. The anisotropy of YCo_4B, at ambient pressure, changes from uniaxial to planar one, at T_{SR} = 159 K [5]. As effect of pressure, the T_{SR} temperature decreases with a rate of 22 K/GPa [26] and thus, at pressures estimated for the isomorphous structure transition (p > 7 GPa), the uniaxial anisotropy is still present, and consequently, a possible spin reorientation to planar anisotropy.

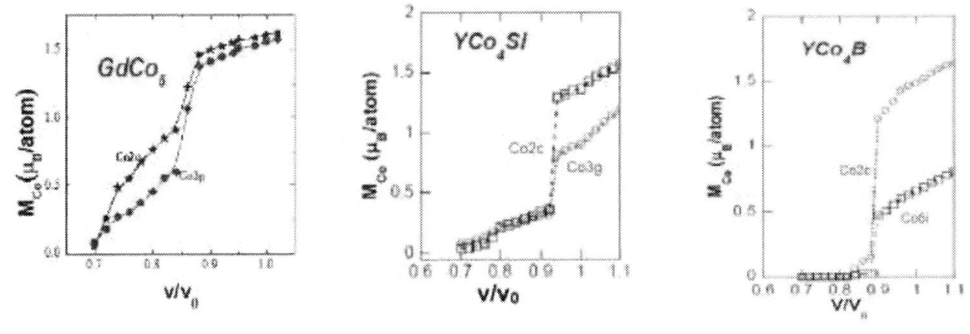

Figure 2. The dependences of the cobalt moments, at 2c and 3g/6i sites, on relative volumes, describing the pressure effects for $GdCo_5$, YCo_4Si and YCo_4B compounds.

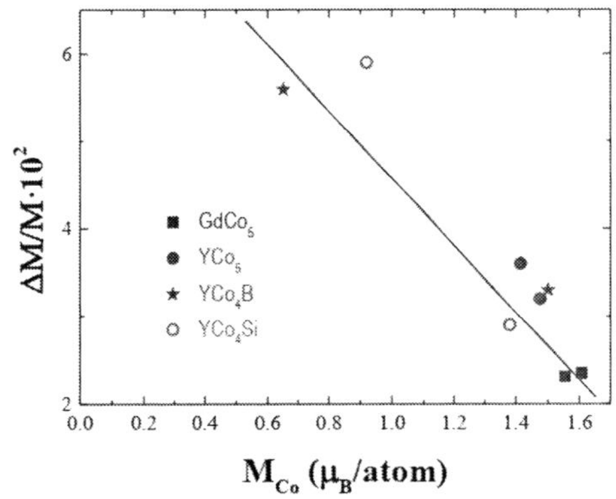

Figure 3. The rate of variations of cobalt moments at 2c and 3g/6i sites, in hexagonal compounds in the range $0.965 \le v/v_0 \le 1.0$, as function of their values at v/v_0=1.

The Gd5d and Y5d band polarizations follow, as function of pressure, a similar trend as the cobalt moments. The induced polarization by 4f-5d exchange is of ≅ 0.19 μ_B, as determined in $GdCo_5$, at relative volume (pressure), v/v_0 = 0.7, when the cobalt magnetic moment is nil. The induced polarizations, M_{5d}(d) and M_{4d}, by 5d(4d)-3d short range exchange interactions are proportional to the number of magnetic atoms situated in the first coordination shell to an R site, z_i and their magnetic moments, M_i. The ratio $|M_{5d}(d)/M_{4d}|/\sum z_i M_i = 1.25\times10^{-2}$. The same value has been already evidenced in RM_5 (M = Co, Ni), compounds [7, 10], being a characteristic feature of all hexagonal RCo_5-based compounds.

The exchange interactions, at the level of unit cell, are mediated by R5d/Y4d band polarizations. The changes of magnetic behaviour, of R and Co sublattices, as well as the implied exchange interactions, are mirrored by the R5d/Y4d band polarizations. This is also confirmed by linear dependence of the Curie temperatures on R5d/Y4d band polarizations of RCo_5-based samples with different substitutions both at R or Co sites – Fig.4.

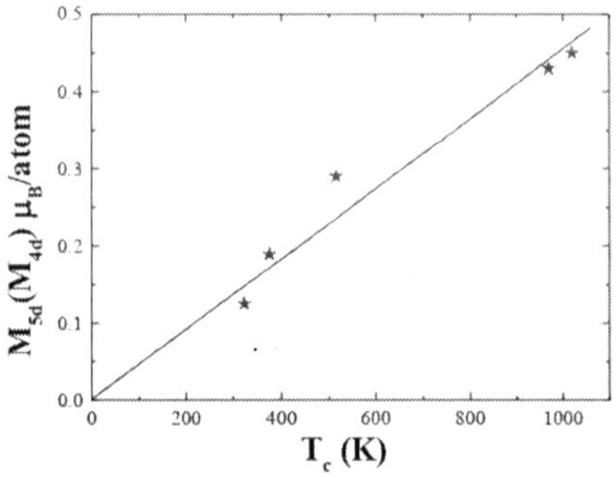

Figure 4. The dependence of Curie temperatures on Gd5d and Y4d band polarizations in some RCo_5 based compounds.

The magnetic properties of ferrimagnetic $Gd_xY_{1-x}Co_4A$ with A = Si or B are analysed also by using a two sublattices mean field model and the average cobalt moments compared with those obtained from band structures. The magnetizations of gadolinium and cobalt sublattices, at 4 K, compensate at x = 0.57 for A=Si and x = 0.43 when A = B – Fig.5. Assuming that gadolinium moment is 7 μ_B, the mean cobalt moments were determined as well the parameters characterizing the interactions inside and between magnetic sublattices.

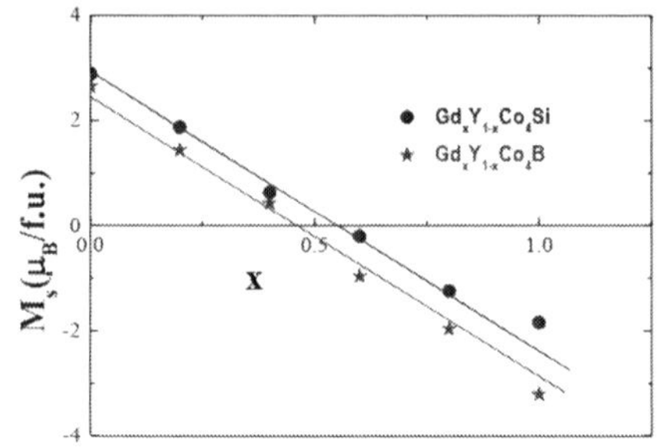

Figure 5. Composition dependences of the magnetizations, at 4K, in $Gd_xY_{1-x}Co_4A$ compounds with A = Si, B.

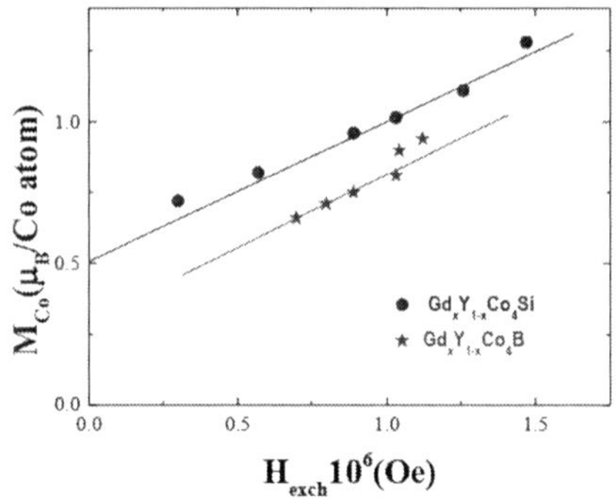

Figure 6. The dependence of the mean cobalt moments on the exchange field in $Gd_xY_{1-x}Co_4A$ series with A=Si, B.

A linear dependence of mean cobalt moments on the exchange field, H_{ex} is evidenced – Fig. 6 – with a rate $\Delta M_{Co}/\Delta H_{exch} \cong (3\cdot10^6)^{-1}$ μ_B/Oe, similar as that reported in RCo_2 based compounds [27]. The mean induced cobalt moments in the composition range $0 \leq x \leq 1.0$ are of $\cong$ 0.52 μ_B (A = Si), 0.2 μ_B (A = B) and of 0.1 μ_B when A = Co. These values are close to those determined from band structure, when considering the mean cobalt moments at 2c and 3g/6i sites: (0.57, 0.44 μ_B) for A = Si, (0.2, 0.17 μ_B)

when A = B and (0.17, 0.06 μ_B) if A = Co. These data indicate that the exchange splitting of Co3d bands is nearly proportional to the internal field intensities.

As effect of pressure, or through the variations of the internal fields, the exchange splitting energies of Co3d bands and the related cobalt moments are modified. The mean field model, gives only the general trend on the evolution of cobalt moments, the band structure calculations allowing a more accurate description of magnetic moments and exchange interactions in connection with the considered cobalt sites.

The exchange interactions, in RCo_5 based compounds, at the level of the unit cell, are mediated by R5d band polarizations. The Y4d band polarizations mirror all changes in the cobalt sublattice magnetizations. Thus, the M_{5d} and M_{4d} values are key points in describing the exchange interactions in a considered system, as also confirmed by their linear trend when plotting with respect to Curie temperatures.

4. Conclusions

The pressure dependences of the cobalt moments in hexagonal compounds, evidence the presence of a first order isomorphous structure transitions, parallelly with the cobalt moment collapse from high spin state to low spin one. The Fermi surface topology, at the transition, is also changed. The R5d and Y4d band polarizations mirror the changes in the magnetic properties as effect of substitutions both at R and Co sites, mediating also the exchange interactions. The mean field model can describe only the general features of the magnetic properties of the hexagonal compounds, but not at the atomic level particularly when non-equivalent sites are present in the considered structure.

References

[1]. E.Burzo, A.Chelkowoski, H.R. Kirchmayr, Landolt Börnstein Handbook, Springer Verlag, Vol.19d2, 1990
[2]. K. H. J. Buschow, Handbook of Magnetic Materials, 10 463 (1997)
[3]. A. T. Pedziwiatr, S. Y. Jiang, W. E. Wallace, J. Magn. Magn. Mater. 66, 69 (1987)
[4]. R. L. Streever, Phys. Rev. B19, 2704 (1979)
[5]. T. T. Dung, N. P. Thuy, N. M. Hong, T. D. Hien, Phys. Stat. Solidi (a) 106, 201 (1988)
[6]. I. A. Campbell, J. Phys. F: Metal Phys. 2, L47 (1972)
[7]. E. Burzo, L. Chioncel, R.Tetean, O. Isnard, J. Phys.: Condens. Matter, 23, 026001(2011)
[8]. E. Burzo, L. Chioncel, I. Costina, S. G. Chiuzbaian, J. Phys.: Condens. Matter, 18, 4861 (2006)
[9]. L. Néel, Ann. Phys. (Paris) 3, 147 (1948)
[10]. E. Burzo, P. Vlaic, I. Creangă, J. Alloys Comp. 509, 3289 (2011)
[11]. E. Burzo, I. Creangă, M. Ursu, Solid State Commun. 64, 586 (1987)
[12]. O. K. Andersen, Phys. Rev. B 22, 3060 (1975)
[13]. O. K. Andersen, O. Jepsen, Phys. Rev. Letters, 55, 2571 (1984)
[14]. U von Barth, L. Hedin, J. Phys. C.: Solid State Phys. 5, 1629 (1972)
[15]. C. Zlotea, O. Isnard, J. Magn. Magn. Mater. 253, 118 (2002)
[16]. T. Turek, C. Becker, J. Hafner, J. Phys.: Condens. Matter, 4, 7257 (1992)
[17]. E. Burzo, P. Vlaic, AIP Conf. Proc. 103-110, 1564 (2013)
[18]. D. Koudela, U. Schwarz, H. Rosner, U. Bukhardt, A. Handstein, M. Hanfland, M. D. Kuzmin, I. Opahle, K. Koepernik, K. H. Müller, M. Richter, Phys. Rev. B 77, 024411 (2007)
[19]. I. M. Lifsitz, Sov. Phys. JETP 11, 1130 (1960)
[20]. E. Burzo, P. Vlaic, D. P. Kozlenko, S. E.. Kichanov, N. O. Golosova, B. N. Savenko, be published
[21]. L. Nordström, M. S. S. Brooks, B. Johansson, J. Phys.: Condens. Matter, 4, 3261 (1992)
[22]. M. M. Yamaguchi, S. Asano, J. Appl. Phys. 79, 5952 (1996)
[23]. H. J. F. Jansen, Phys. Rev. B 59, 4699 (1999)

[24]. H. Yamaoka, Y. Yamamoto, E. F. Schwier, N. Tsujii, M. Yoshida, Y. Ohta, H. Sakurai, J. F. Lin, N. Hiraoka, H. Ishii, K. D. Tsuei, M. Arita, K. Shimada, J. Mizuki Phys. Rev. B 94, 165456 (2016)
[25]. L. Steinbeck, M. Richter, H. Eschrig, Phys. Rev. B 63, 184431 (2001)
[26]. Z. Arnold, J. Kamarád, Y. Skorokhod, N.M. Hong, N.P. Thuy, C.V. Thang, J. Magn. Magn. Mater. 262, 382 (2003)
[27]. E. Burzo, Solid State Commun. 14, 1295 (1974)

Crystalisation of aqueous ferrofluids at the free liquid interface investigated by specular and off-specular x-ray reflectometry

I V Gapon[1,2,3*], V I Petrenko[1,2], O Soltwedel[4,5], Yu N Khaydukov[4], M Kubovcikova[6] P Kopcansky[6], L A Bulavin[2], M V Avdeev[1]

[1] Frank Laboratory of Neutron Physics, Joint Institute for Nuclear Research, Dubna, Russia
[2] Taras Shevchenko National University of Kyiv, Kyiv, Ukraine
[3] Institute for Safety Problems of Nuclear Power Plants, NAS Ukraine, Kyiv, Ukraine
[4] Max-Planck-Institute for Solid State Research, Outstation at MLZ, Garching, Germany
[5] Technical University of Munich, Garching, Germany
[6] Institute of Experimental Physics, SAS, Kosice, Slovakia

E-mail: gapon@jinr.ru

Abstract. Structural organization of nanoparticles from aqueous ferrofluids on free liquid surface was studied by X-ray reflectometry. The observed layered structure at interface is associated with the evaporation of the solvent. By orienting an external magnetic during evaporation of the aqueos ferrofluids their structural organization can be manipulated. For a magnetic field applied perpendicular to the surface a more pronounced ordering along the surface normal is observed as in the case of a parallel field. Independent on the orientation of the magantic field a $\sim 20\ \mu m$ thick surface layer of depleted nanoparticle concentration is found at the interface.

1. Introduction

Ferrofluids (FFs) are fine stable dispersions of magnetic nanoparticles in liquids and belong to the class of advanced nanomaterials. To ensure the long-term stability of FFs the magnetic nanoparticles are coated with some stabilization layers, which prevent the particle coagulation induced by attractive van der Waals forces and magnetic interactions. FFs have been widely used in various industrial and technical fields. The present-day interest to these systems is also associated with their prospects in actively developing medical applications such as drug targeting, magnetic resonance imaging (contrast medium), magnetic hyperthermia for cancer treatment, etc. [1, 2]. For this purpose, many efforts are concentrated to the synthesis of highly stable, reproducible and controllable water-based ferrofluids. Structure characterization of FFs in last two decades focused mainly on the organization of bulk solutions, which is well accessible via small-angle X-ray (SAXS) and neutron (SANS) scattering (e.g., see review [3, 4]). At the same time, because of specific adsorption properties, the behavior of magnetic nanoparticles and surfactants in FFs at interfaces with solids and gases differs significantly from that in bulk, which is to be also considered with respect to application purposes [5–7]. An additional point is that a possible difference in the stability of FFs in bulk and at interfaces should be taken into account in the requirements for the stabilization of these systems with respect to their storage, namely their destabilization at hydrophilic and hydrophobic interfaces like functionalized container walls or the free liquid surface under varying storing conditions (temperature, humidity etc.).

In our previous works [8–11] we investigated the structure organization of ferrofluids at a silicon wall by neutron reflectometry. We found that a single layer of sodium oleate stabilized FFs, dissolved in polar as well as non-polar solvents, arranges at the interface as a transition zone [8, 10]. Additionally we proved that the specific adsorption properties of ferrofluids at interfaces is independent on their core material but can be manipulated by the stabilizing shell [11]. According to neutron reflectometry and small-angle neutron scattering experiments we found a clear correlation between the bulk and its structure at the solidliquid interface for initial aqueous ferrofluids and after PEG polymer modification of water-based FFs[9].

Recently, several works were published in which behavior of ferrofluids at interface with air is investigated. Thus, in [6, 7] formation of the specific layers from surfactant and magnetic particles on free surface of aqueous FFs was reported. Such a structural organization was observed irrespective of the concentration of particles (volume fraction of magnetite 2% and 7%) and after repeated cleaning of the surface. Therefore, the authors concluded [6] that this structure is a characteristic structural organization of magnetic particles near the free surface for aqueous FFs with steric stabilization. This result was verified in [7]. In order to obtain a more homogeneous and controlled surface of the FFs, the authors attempted to exclude the specific layered organization of FFs components in two different ways. However, in both cases, such a structural organization was observed on the surface, but with a slightly different concentration of particles and the thickness of the layer. At the same time, the dependence of this effect on the concentration of magnetite in volume was not detected. The authors concluded [7] that the described two-layer material can be assumed as an essential equilibrium feature inevitably existing at the interface of water-based ferrofluids with gas. For organic FFs only one layer of magnetic nanoparticles was observed [6].

In the present work we investigate the free surface of ferrofluids using X-ray reflectometry. From the data obtained, we can conclude that the self-organization of FFs nanoparticles at the interface with gas is depended on concentration of nanoparticles and the direction of the applied external magnetic field. An additional particles organization near the surface is observed for perpendicular magnetic field.

2. Experimental

Preparation of magnetic fluid was based on the co-precipitation reaction to produce nanomagnetite (Fe_3O_4). The FFs were synthesized at the Institute of Experimental Physics, Slovak Academy of Sciences (IEP SAS), Slovakia, in accordance with the procedure described in [12]. To avoid aggregation between magnetite particles they were covered by a double layer of sodium oleate ($C_{17}H_{33}COONa$) with a surfactant/magnetite mass ratio ranging from 0.73 to 1. Two FFs with magnetite volume fractions of $\varphi_m = 0.4\ vol.\ \%$ and $\varphi_m = 6.6\ vol.\ \%$ were prepared. The exact magnetite concentrations in the FFs were determined by the static magnetization analysis (MPMS SQUID magnetometer, IEP SAS).

The free surface of FFs was investigated at the NREX instrument (Garching, Germany) using the X-ray reflectometer mode. The scans were either done in $\theta - \theta$ geometry, i.e. the sample position was fixed and only detector and source were moved or a series of detector scans for fixed incidence angles to map the reciprocal space. In both cases the sample was not moved during the experiment. To study the surface of the the liquid system in a reflectometer, a bath of inert material is used. To reduce the evaporation, the bath is placed in a closed casing, which allows to create additional pressure and maintain a certain temperature and to provide the surroundings of the bath with an inert gas. Fig. 1 shows the details of the experimental set-up. While on the left side the magnetic field is applied vertically with respect to the liquid interface it is parallel oriented on the right picture. Between the steel plates, which are separated by

IOP Conf. Series: Journal of Physics: Conf. Series **994** (2018) 012008 doi:10.1088/1742-6596/994/1/012008

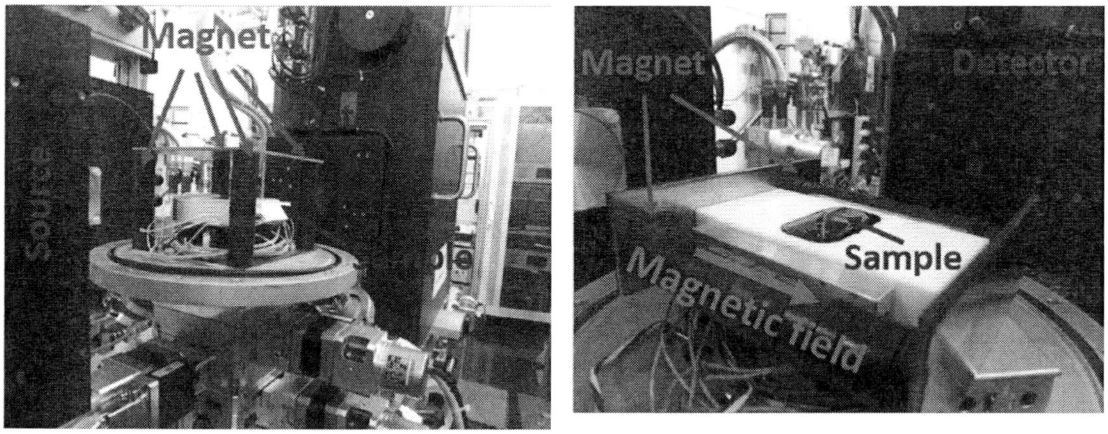

Figure 1. Photographs of the working chamber of the reflectometer NREX for the study of liquid systems using X-rays.

permanent magnets, a trough made of PTFE is placed. In this setting magnetic fields of 100 Oe at the trough position and in the respective orientations were achieved. . It should be noted that it is necessary to provide a special sample holder for the reduction of any mechanical vibrations during the study of free surfaces of liquid systems. For this purpose, specific anti-vibration platforms are used, which can extinguish even significant fluctuations.

3. Results and discussion

During the measurement of the magnetic fluid with concentration of 0.4 *vol.* % the solvent was evaporating from ferrofluid. For correct interpretation of the data, the reflectometry curves were taken during and after the complete evaporation of the solvent. The data are presented in fig. 2a. Scattering map for this system after solvent evaporation is shown in fig. 2b. Complex structural organization on the free surface of the ferrofluid was concluded from the resulting curve. The

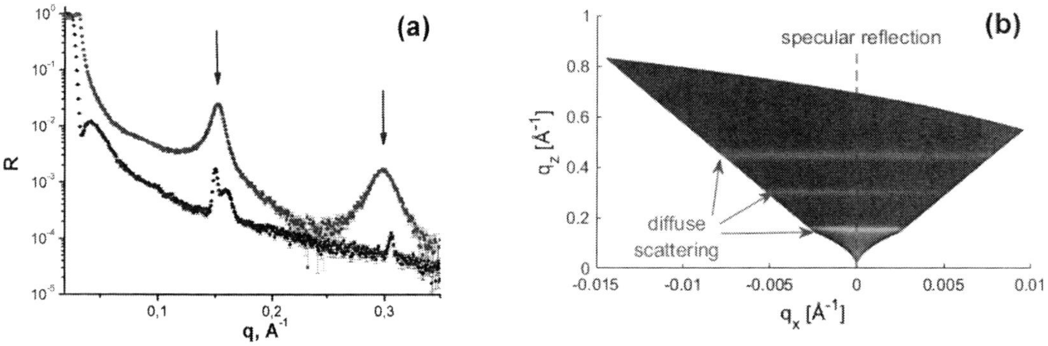

Figure 2. (a) Experimental reflectometry dependences for the free surface of the aqueous ferrofluids before (black circles) and after evaporation of the solvent (red squares). Arrows show correspondent Brag peaks at reflectivity curves. (b) Scattering map for ferrofluids after evaporation of the solvent. Specular reflection is presented by dashed line. Diffuse scattering are shown by arrows.

found Bragg-peaks at reflectivity curves (fig. 2) indicate a periodic structure in both cases. In addition, after the evaporation we investigated the formed layered structure using off-specular reflectivity. Since the evaporation is an ongoing process during a x-ray scan it is hopeless to model the reflectivity curves. Instead we make use of the position of the characteristic Bragg-peaks. According to the rule $D = 2\pi/q$, where is the wave vector transfer between the maxima on the curve. The resulting size is about 4 nm. This is twice less than the average diameter ($D \sim 8\ nm$) of the magnetic particles in the ferrofluid. This means that when the solvent is completely evaporated from a magnetic fluid, the magnetic particles are packed in a layered structure with a dense packing. On reflectometry curves (see fig. 2a), obtained before and after evaporation of the solvent, we see the coincidence of the position of the Bragg-peaks. From this fact, it can be assumed that during the investigation of the free surface of the ferrofluid there is a process of evaporation of the solvent from the near-surface layer of the liquid system, which increases the concentration of nanoparticles in the near-surface layer. An increase in the concentration of particles in this layer leads to their aggregation. Therefore, to exclude effect of solvent evaporation on reflectometry data, special cell for magnetic fluid should be used in future experiments. Form the off-specular measurement (fig. 2b) we have to conclude that the NP arrangement seems to be isotropic in 3D. Since the probed lateral q-space is rather small, additional scans under grazing incidence at larger q-values are necessary to prove this hypothesis.

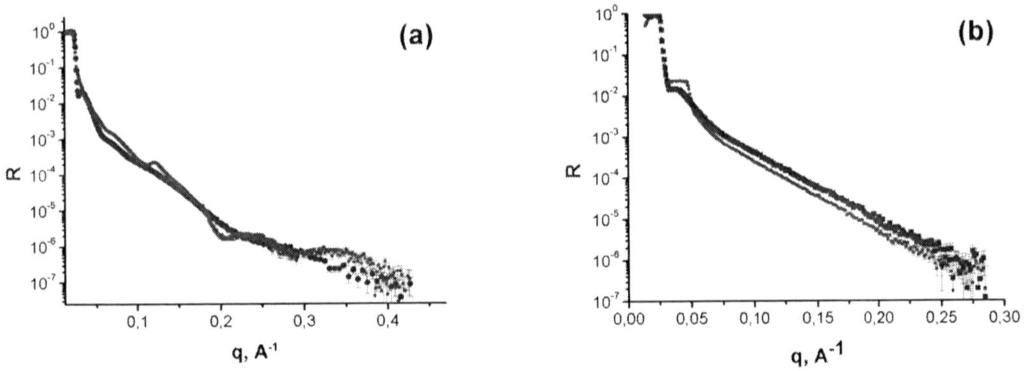

Figure 3. Experimental data of the X-ray reflectometry experiment for the free surface of ferrofluid with a volume fraction of magnetite of 0.4% (a) and 6.6% (b) under the application of the magnetic field in parallel (blue squares) and perpendicularly (red circles) to the surface of the ferrofluids

The additional investigation of the structural reorganization of nanoparticles at the free surface was carried out for two concentrations of magnetite in the ferrofluids under application of an external magnetic field. The experiment was done at two direction of field (perpendicular or parallel to the surface of the ferrofluids) and constant magnetic field with an induction of 0.01 T. Ferrite magnets with two metal plates were used for creation magnetic field at the sample, as is presented on fig. 1. In fig. 3a the reflection as function of the scattering vector q is shown for both magnetic orientations for the free surface of the magnetic fluid with a $\varphi_m = 0.4\ vol.\ \%$. In the investigated system, with the perpendicular direction of the magnetic field, we can see the additional structural organization of the magnetic particles. This corresponds to the creation of a layer of particles near the surface. Obviously the orientation of the magnetic field has an impact on the resulting NP arrangement. Due to technical reasons the liquid sample is instable during the measurements and these results have only preliminary character. Nevertheless for higher NP-concentration ($\varphi_m = 6.6\%$) two critical edges are obsereved for both field directions.

The lower one at $q = 0.023\ \AA^{-1}$ corresponds well with the expected value for pure sodium oleate ($q_{crit,C_{18}H_{33}NaO_2} = 0.0229\ \AA^{-1}$ at $\rho = 1.2\ g/cm^3$) and the higher at $q = 0.047\ \AA^{-1}$ with the one for magnetite $q_{crit,Fe_3O_4} = 0.0444\ \AA^{-1}$ at $\rho = 5.2\ g/cm^3$). The steep step between both plateaus of total external reflection ($\sim$ 97% attenuation) can serve as a rough estimate for the thickness of the absorbing layer. Since the k-absoption of iron ($7.11\ keV$) is close to the one for Cu ($8.14\ keV$) strong x-ray fluorescence is expected. Indeed the mass attenuation coefficient for Fe_3O_4 exceeds for photons with 8 keV a value of $\mu = 2400\ cm^{-1}$. Using the Lambert-Beer absorption law one calculates a thickness of roughly 20 μm.

4. Conclusions

As a result of the work, it was discovered that the evaporation of the solvent from the free surface of the ferrofluid leads to the aggregation of its nanoparticles and the creation of the layered structure. It was found that structural organization of nanoparticles at the air/FFs interface dependent on concentration of magnetite in initial FF. The highest effect on additional particles organization near surface is observed for perpendicular to the surface magnetic field.

Acknowledgments

The work was financed via Russian scientific foundation (project 14-12-00516). Also this work was supported by the VEGA 0045, Radiomag COST TD 1402, MERANET MACOSYS and by Ministry of Education Agency for Structural Funds of EU in frame of Projects 26220120005 and 26220120033.

References

[1] Odenbach S (ed) 2009 *Colloidal Magnetic Fluids* vol 763 (Berlin: Springer-Verlag)
[2] Torres-Daz I and Rinaldi C 2014 *Soft Matter* **10** 8584–8602
[3] Vekas L, Bica D and Avdeev M 2007 *China Part.* **5** 505–511
[4] Avdeev M V and Aksenov V 2007 *Phys. Uspekhi* **53** 505–511
[5] Vorobiev A, Major J, Dosch H, Gordeev G and Orlova D 2004 *Physical Review Letters* **93** 267203
[6] Vorobiev A, Gordeev G, Konovalov O, Gordeev G and Orlova D 2009 *Phys. Rev. E* **79** 031403
[7] Vorobiev A, Konovalov O, Khassanov A *et al.* 2011 *J. Appl. Phys.* **110** 102219
[8] Gapon I, Petrenko V, Avdeev M *et al.* 2015 *J. Surf. Invest.: X-Ray, Synchrotron Neutron Tech.* **9** 320–325
[9] Kubovcikova M, Zavisova V, Gapon I *et al.* 2016 *J. Magn. Magn. Mater.* **427** 67–70
[10] Avdeev M, Petrenko V, Gapon I *et al.* 2017 *Appl. Surf. Sci.* **848** 012015
[11] Gapon I, Petrenko V, Bulavin L *et al.* 2015 *J. Phys. Conf. Ser.* **352** 4953
[12] Zavisova V, Koneracka M, Muckova M *et al.* 2011 *J. Magn. Magn. Mater.* **323** 14081412

Small-angle neutron scattering investigations of Co-doped iron oxide nanoparticles. Preliminary results

Dorina Creanga[1], Maria Balasoiu[2,3*], Dmitro Soloviov[3,4,5], Alexandra-Maria Balasoiu-Gaina[2,6,7], Emil Puscasu[1], Nicoleta Lupu[8], Cristina Stan[9]

[1] "Alexandru Ioan Cuza" University, Physics Faculty, Iasi, Romania
[2] Joint Institute for Nuclear Research, Dubna, Russia
[3] "Horia Hulubei" Institute of Physics and Nuclear Engineering, Bucharest, Romania
[4] Moscow Technical Physics Institute, Dolgoprudnyi, Russia
[5] Institute for Safety Problems of Nuclear Power Plants of the Ukrainian NAS, Kyiv, Ukraine
[6] CMCF, Moscow State University, Moscow, Russian Federation
[7] West University of Timisoara, Timisoara, Romania
[8] National Institute of Research and Development for Technical Physics, Iasi, Romania [9] University "Politehnica" of Bucharest, Faculty of Applied Sciences, Physics Department, Romania

E-mail: balas@jinr.ru

Abstract. Preliminary small-angle neutron scattering investigations on aqueous suspensions of several cobalt doped ferrites ($Co_xFe_{3-x}O_4$, x=0; 0.5; 1) nanoparticles prepared by chemical co-precipitation method, are reported. The measurements were accomplished at the YuMO instrument in function at the IBR-2 reactor. Results of intermediary data treatment are presented and discussed.

1. Introduction

Cobalt ferrite nanoparticles ($CoFe_2O_4$) represent promising materials for technological and medical applications [1-15]. Active works have been reported in connection with new developments in preparation methods [16-29] and properties investigation of resulted samples [30-36].

In the present paper preliminary small-angle neutron scattering investigations on aqueous suspensions of several cobalt doped ferrites ($Co_xFe_{3-x}O_4$, x=0; 0.5; 1) nanoparticles prepared by chemical co-precipitation method, are reported.

The preparation of $Co_xFe_{3-x}O_4$ with different Co doping levels (x=0; 0.25; 0.5; 0.75; 1) using the chemical co-precipitation method, was reported earlier [37] and the scanning electron microscopy analysis of the samples showed that the cobalt ferrite nanoparticles have spherical shapes and average sizes of 70 nm [37].

2. Experimental

Samples of $Co_xFe_{3-x}O_4$ with different Co doping ratio (x=0; 0.5; 1) were prepared and preliminary small angle neutron scattering measurements performed.

2.1. Sample preparation

The investigated cobalt ferrite nanoparticles were prepared by a chemical route based on highly versatile and of low cost Massart's method [38]. The iron compound source was the ferric chloride hydrated crystalline powders and the ferrous sulfate while the cobalt ions were provided by the cobalt sulfate. The high purity reagents were purchased from Sigma-Aldrich firm and used without additional purification.

Alkali synthesis medium was ensured based on hot 24% sodium hydroxide pouring under continuous stirring, while the stabilization of nanoparticles in aqueous suspension was accomplished with citric acid as capping agent (5 g citric acid in 10 ml deionized water for each about 4 g of ferrophase sample).

2.2. SANS experiment

Small angle neutron scattering (SANS) experiments were performed at the time-of-flight YuMO spectrometer [39] in function at the high flux pulse IBR-2 reactor, JINR Dubna. The Sonix+ software control system provides spectrometer operation [40]. The experiments were carried out at a sample-to-detector distances of 5.28 m and 13.04 m, resulting in a Q range of 0.007÷0.2 A^{-1}. The sample diameter and thickness in the beam were 14 mm and respectively 1 mm. The measured neutron scattering spectra are corrected relative to the transmission and thickness of the sample, background scattering due to the experimental cuvette and the vanadium reference sample using the SAS software [41], providing a neutron scattering intensity in absolute units of cm^{-1}.

3. Results and discussion

In Figure 1 the experimental curves of small-angle neutron scattering from the aqueous solutions of three type of nanoparticles, i.e. Fe3O4, $Co_{0.5}Fe_{2.5}O_4$ and $CoFe_2O_4$, are given.

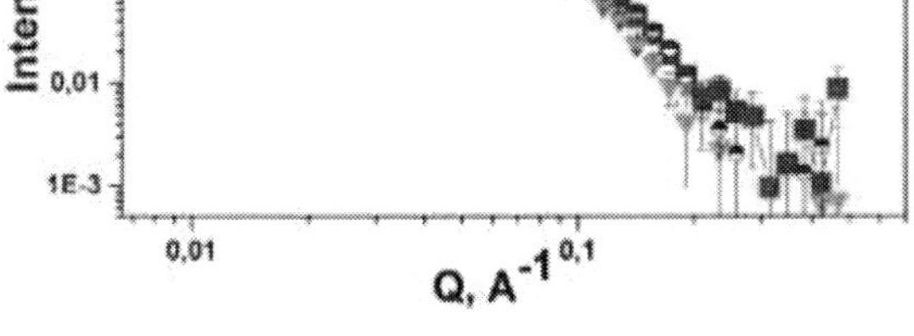

Figure 1. Small-angle neutron scattering experimental curves from aqueous solutions of nanoparticles: (a) Fe_3O_4 (red squares); (b) $Co_{0.5}Fe_{2.5}O_4$ (half-blue circles); (c) $CoFe_2O_4$ (green triangles).

The experimental data from the Q range of (0.007÷0.07) $Å^{-1}$ were preliminary fitted with the Programme FITTER [42] in the monodisperse approximation. Two axis ellipsoidal coreshell for Fe_3O_4 (a) and triaxial ellipsoidal coreshell for $Co_{0.5}Fe_{2.5}O_4$ (b) and $CoFe_2O_4$ (c) models, respectively, were obtained (Figure 2).

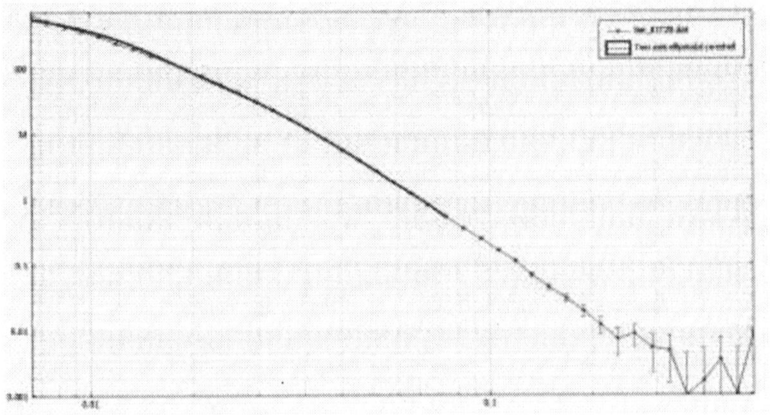

a)

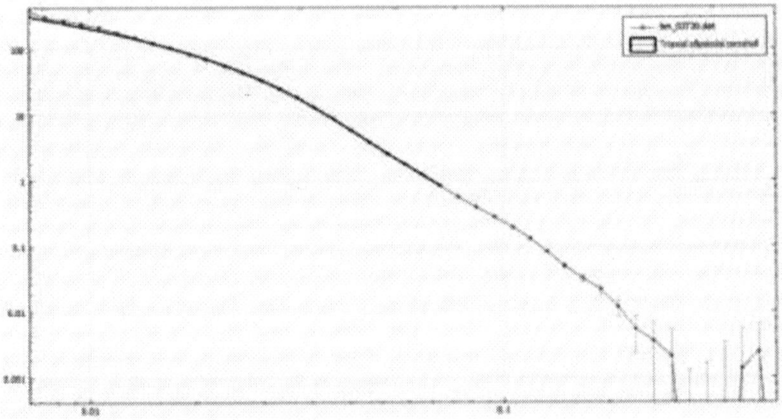

b)

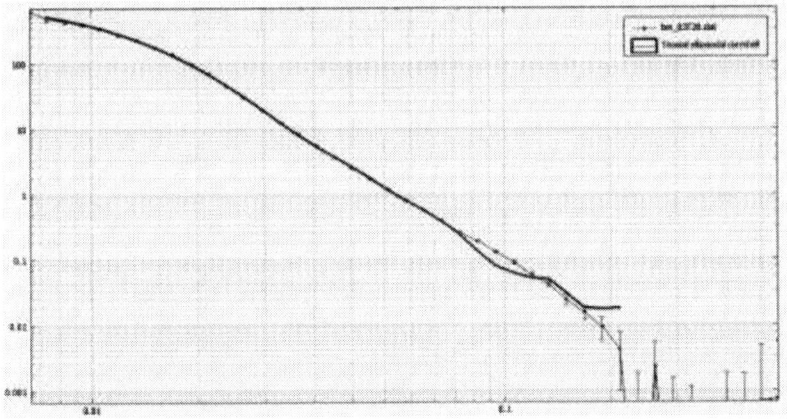

c)

Figure 2. Model fit of experimental SANS curves in the Q range of (0.007÷0.07) Å^{-1} using FITTER programme: (a) for Fe_3O_4 nanoparticles; (b) for $Co_{0.5}Fe_{2.5}O_4$; (c) for $CoFe_2O_4$.

The model parameters resulting from the data fitting are given in the Table 1.

Table 1. The results of the SANS experimental data model fitting in the Q range of (0.007÷0.07) $Å^{-1}$.

Sample	**Model fit**	**Semiaxis of core [nm]**	**Thickness of shell [nm]**
Fe_3O_4	Two axis ellipsoidal coreshell	21.1±0.1 ε=0.0054, eccentricity	9.1±0.1
$Co_{0.5}Fe_{2.5}O_4$	Triaxial ellipsoidal coreshell	a=38.8±0.1 b=3.3±0.1 c=5.4±0.1	6.3±0.1
$CoFe_2O_4$	Triaxial ellipsoidal coreshell	a=21.2±0.1 b=6.1±0.1 c=9.9±0.1	10.5±0.1

4. Conclusions

Preliminary small-angle neutron scattering investigations on aqueous suspensions of several cobalt doped ferrites ($Co_xFe_{3-x}O_4$, x=0; 0.5; 1) nanoparticles prepared by chemical co-precipitation method, were accomplished.

Small-angle neutron scattering measurements fitted in the Q range of (0.007÷0.07) $Å^{-1}$ showed for all the samples ellipsoidal coreshell model of 20÷40 nm. A difference between the scanning electron microscopy (SEM) and SANS results was determined. In the case of SEM, the particles could be aggregated and this fact can explain the obtained double value for the particle size, in comparison with those measured by SANS.

Further, the experimental data will be fitted in the next Q domain, (0.07÷0.02) $Å^{-1}$ and the polydispersity will be considered.

Acknowledgements. JINR-RO Projects 219/10.04.2017 and 220/10.04.2017 item 62 and JINR-RO Projects for 2018 year are acknowledged.

References

[1] Shenker H 1957 Phys.Rev. **107** 1246
[2] Jordan A, Wust P, Fahling H, John W, Hinz A and Felix R 1993 Int. J. Hyperthermia **9** 51
[3] Kitamoto Y, Kantake S, Shirasaki S, Abe F, Naoe M 1999 J. Appl. Phys. **85** 4708
[4] Ngo A-T and Pileni M-P 2000 Adv. Mater. **12**(4) 276
[5] Fortin J P, Wilhelm C, Servais J, Menager C, Bacri J C, Gazeau F 2007 J. Am. Chem. Soc. **129** 2628
[6] Kim D H, Nikles D E, Johnson D T, Brazel C S 2008 J. Magn. Magn. Mater. **320** 2390
[7] Creangă D E, Iacob Gh, Ursache M, Nădejde C, Răcuciu M 2008 J Optoelectron Adv Mater. **10**(3) 628
[8] Zi Z, Sun Y, Zhu X, Yang Z, Dai J, Song W 2009 J. Magn. Magn. Mater. **321** 1251
[9] Gao J, Gu H, Xu B 2009 Acc. Chem. Res. **42**(8) 1097
[10] Kami D, Takeda S, Itakura Y, Gojo S, Watanabe M, Toyoda M 2011 Int. J. Mol. Sci. **12** 3705

[11] Zheng J, Song X, Liu X, Chen W, Li Y, Guo J 2012 Materials Letters **73** 143
[12] Kaur Rajput J, Kaur G 2013 Chinese Journal of Catalysis **34** 1697
[13] Drašler B, Drobne D, Novak S, Valant J, Boljte S, Otrin L, Rappolt M, Sartori B, Iglič A, Kralj-Iglič V, Šuštar V, Makovec D, Gyergyek S., Hočevar M, Godec M, Zupanc 2014 J, Int. J. Nanomedicine **9** 1559
[14] Oprica L, Nadejde C, Andries M, Puscasu E, Creanga D, Balasoiu M 2015 Environ. Eng. Manag. J. **14**(3) 581
[15] Sergeenkov S, Stan C, Cristescu C P, Balasoiu M, Perov N S, Furtado C 2017 Philosophical Magazine Letters **97**(7) 287
[16] Puscasu E, Sacarescu, L., Lupu, N. N. Lupu, M. Grigoras, Oanca G, Balasoiu M, Creanga D 2016
J Sol-Gel Sci Technol **79** 457(9)
[17] Cheng F, Peng Z, Liao C, Xu Z, Gao S, Yan C, Wang D, Wang J 1998 Solid. State. Commun. **107**(9) 471
[18] Tourinho F A, Franck R, Massart R, Perzynski R 1989 Prog. Colloid. Polym. Sci. **79** 128
[19] Charles S W, Chandrasekhar R, O'Grady K, Walker M 1988 J. Appl. Phys. **64**(10) 5840
[20] Morais P C, Garg V K, Oliveira A C, Silva L P, Azevedo R B, Silva A M L, Lima E C D 2001 J. Magn. Magn. Mater. **225** 37
[21] Jiao Z, Geng X, Wu M, Jiang Y, Zhao B, 2008 Colloids Surf. A: Physicochem. Eng. Aspects **313-314** 31
[22] Artus M, Ammar S, Sicard L, Piquemal J Y, Herbst F, Vaulay M J, Fievet F, Richard V 2008 Chem. Mater. **20** 4861
[23] Babic-Stojic B, Jokanovic V, Milivojevic D, Jaglicic Z, Makovec D, Jovic N, Marinovic-Cincovic M 2013 Journal of Nanomaterials 741036(9)
[24] Covaliu C I, Matei C, Ianculescu A, Jitaru I, Berger D 2009 U.P.B. Sci. Bull., Series B 71(4) 53
[25] Moumen N, Pileni M P 1996 Chem. Mater. **8** 1128
[26] Tartaj P, Del Puerto Morales M, Veintemillas–Verdaguer S, Gonzales–Carreno T, Serna C J 2003 J. Phys. D: Appl. Phys. **13** R182
[27] Bao N, Shen L, Padhan P, Gupta A 2009 Appl. Phys. Lett. **92** 173101(3)
[28] Herrera A P, Polo-Corrales L, Chavez E, Cabarcas-Bolivar J, Uwakweh O N C, Rinaldi C 2013 J. Magn. Magn. Mater. **328** 41
[29] Zhao L, Zhang H, Xing Y, Song S, Yu S, Shi W, Guo X, Yang J, Lei Y, Caot F al. 2008 J. Solid State Chem. **181** 245
[30] M. Balasoiu, O. I. Ivankov, D.V. Soloviov, S.N. Lysenko, R.M. Yakushev, A-M. Balasoiu-Gaina, N. Lupu 2015 J. Optoelectron. Adv. Mater. **17**(7-8) 1114
[31] T. N. Mamedov, D.S. Andrievskii, M. Balasoiu, K.I. Gritsaj, V.N. Duginov, A.L. Getalov, E.N. Komarov, S.A. Kotov, A.E. Moroslip, I.I. Pavlova, V.G. Scherbakov, S.I. Vorob'ev, C. Stan 2015 J. Optoelectron. Adv. Mater. **17**(7-8) 1086
[32] Avdeev M V, Petrenko V I, Gapon I V, Bulavin L A, Vorobiev A A, Soltwedel O, Balasoiu M, Vekas L, Zavisova V, Kopcansky P 2015 Applied Surface Science **352** 49
[33] Balasoiu M, Kuklin A I 2016 Rom. Journ. Phys. **61**(3–4) 473
[34] Puscasu E, Sacarescu L, Domocos A, Leostean C, Turcu R, Creanga D, Balasoiu M 2016 Romanian Journal of Physics **61**(5–6) 946
[35] I.V. Gapon, V.I. Petrenko, L.A. Bulavin, M. Balasoiu, M. Kubovcikova, V. Zavisova, M. Koneracka, P. Kopcansky, M.V. Avdeev 2017 Journal of Physics: Conf. Series **848** 012015(5)
[36] Balasoiu M, Ivankov O, Soloviov D, Lysenko S, Balasoiu-Gaina A-M, Stan C, Lupu N, Creanga D, Kuklin A 2017 Journal of Physics: Conf. Series 848 012026(6)
[37] Tiriba G, Balasoiu M, Puscasu E, Sacarescu L, Stan C, Creanga D E 2017 U.P.B. Sci. Bull. Series A 79(4) 327
[38] Massart R 1981 IEEE Trans. Magn. **17**(2) 1247

[39] Kuklin A I, Islamov A K, Gordeliy V I 2005 Neutron News **16**(3) 16
[40] Kirilov A S, Litvinenko E I, Astakhova N V, Murashkevich S M, Petukhova T B, Yudin V E, Gordelii V I, Islamov A Kh, Kuklin A I 2004 Instruments and Experimental Techniques **47**(3) 3
[41] Soloviev A G, Solovieva T M, Stadnik A V, Islamov A Kh, Kuklin A I 2003 SAS The Package for Small Angle Neutron Scattering Data treatment. Version 2.4. Long write-Up and User's Guide (Russian) JINR Communication P10-2003 86; http://wwwinfo.jinr.ru/programs/jinrlib/sas/indexe.html

[42] Soloviev A G, Stadnik A V, Islamov A H and A.I.Kuklin A I 2008 Fitter. The package for fitting a chosen theoretical multi-parameter function through a set of data points. Application to experimental data of the YuMO spectrometer. Version 2.1.0. Long Write-Up and User's Guide. Communication of JINR E10-2008-2; http://wwwinfo.jinr.ru/programs/jinrlib/fitter/docs/html/node2.html

Magneto-optical properties of binar ferrocolloids

A F Pshenichnikov[1], A V Lebedev[1], E V Lakhtina[1], G V Stepanov[2]

[1] Institute of Continuous Media Mechanics Ural Branch of Russian Academy of Sciences the affiliate of Perm Federal Scientific Research Center, 614013, Perm, Ak. Korolev str., 1, Russia.
[2] State Scientific Center of the Russian Federation "State Research Institute for Chemistry and Technology of Organoelement Compounds", 111123, Moscow, Shosse Entuziastov, 38, Russia

E-mail: pshenichnikov@icmm.ru

Abstract. In this work, a new method for increasing optical anisotropy of a ferrocolloid through introducing the coiled polymer molecules or elongated nanosized non-magnetic particles is realized. Since the dimensions of structural elements comprising such a binary colloidal solution are small compared to the wavelength, the ferrocolloid remains optically homogeneous. Type I binary ferrocolloids are obtained by introducing polybutadiene molecules into a magnetic fluid (magnetite + kerosene + oleic acid). In this case, an increase in the double refraction (DR) is due to the deformation and stretching of the polymer coils along the magnetic field. In weak fields, double amplification of the signal was detected for the concentration of polymer molecules of about 0.5 %. A further increase in the concentration of impurity molecules weakens DR due to a disturbance of the sedimentation stability of the solution and precipitation of colloidal particles. Type II binary solution is synthesized on the basis of a magnetic fluid and rod-shaped impurity nanoparticles of goethite ($\alpha-$FeOOH). The transverse dimension of the impurity particles (10 − 30 nm) was close to the average diameter of single-domain magnetite particles, and the longitudinal dimension was an order of magnitude larger. An increase in the DR occurs due to the orientation of long axes of impurity particles along the magnetic field caused by the difference in the "demagnetizing" coefficients along and across the axis of the particle. The magnetic double refraction has been studied depending on the concentration of magnetite and impurity particles and the strength of the magnetic field. For the first time, an experimental substantiation of the multiple amplification of the DR signal by impurity particles was obtained. In the fields (up to 10 kA/m) and for the volume fraction of impurity particles of the order of one percent, the DR signal is amplified by more than an order of magnitude. In stronger fields, the signal gain, associated with the influence of impurity particles, reaches saturation and, with further increase in the field strength, remains practically unchanged, while the total anisotropy of the solution continues to increase due to the orientation of the magnetite particles.

1. Introduction

Magnetic fluids, which are stable colloid suspensions of single-domain ferromagnetic nanoparticles with a characteristic dimension of $\sim$ 10 nm, in the absence of the external magnetic field are optically homogeneous. However, when exposed to the external field they acquire properties of uniaxial crystal with very strong optical anisotropy. An abnormally large value of the Cotton-Mouton constant is the special feature that arouses keen researchers' interest in the double refraction (birefringence) in magnetic fluids. This effect was investigated in a series of theoretical and experimental works [1]-[12]. Despite a high value of the Cotton-Mouton constant the integral effect of double refraction (DR) in magnetic fluids was found to be rather weak, since due to a strong absorption of light, it is common practice to use in experiments either dilute solutions or very thin (a few tenths or hundredths of a millimeter) layers of magnetic fluid.

In [13] the authors describe a new method of increasing optical anisotropy of a ferrocolloid by introducing the coiled molecules of a polymer into a suspension and present preliminary results. Our work is an extension of the above study. Furthermore, it interprets the results of experiment on the double refraction in the Type II binary solution, containing impurity particles in the form of elongated nonmagnetic particles. Since the dimensions of structural elements comprising such binary colloid solutions are small compared to the wavelength, the ferrocolloid remains optically homogeneous. Hence, the focus of our study is the magnetic and optical properties of new magnetizable media binary ferrocolloids, which differ from the ordinary magnetic fluids by high values of the Cotton-Mouton constant.

2. Type I binary ferrocolloids

In all experiments, a magnetic fluid composed of magnetite particles dispersed in kerosene and stabilized with oleic acid was used as a basis. It was obtained by a standard method of chemical precipitation and subsequent removal of non-magnetic particles and free stabilizer through repeptization. Type I binary ferrocolloids were obtained by introducing polybutadiene molecules into the magnetic fluid. An increase in the optical anisotropy of the solution was achieved due to deformation and stretching of polymer coils along the magnetic field. The main difficulty in obtaining binary colloids is associated with a strong influence of polymer molecules on the aggregative stability of the solution.

To obtain information on the stability of magnetic liquids and binary solutions against aggregation, the dynamic susceptibility of solutions was measured and cluster analysis was carried out based on the expansion of the dynamic susceptibility into a series of the Debye functions and the known relationship between the relaxation time of the magnetic moments of particles (or clusters) and their volume. The method of cluster analysis is described in detail in [14]-[16]. It was found that in magnetic fluids, even if they do not contain polymer molecules, there are a great number of quasispherical aggregates with characteristic dimensions ranging from 40 to 80 nm. The results obtained are in good agreement with the data of previous papers [14]-[16], in which the same magnetic fluids were investigated. As an example, Figs. 1 and 2 show typical dispersion curves for the dynamic susceptibility of a magnetic fluid and the temperature dependence of the average (hydrodynamic) diameter of aggregates. Clearly, the average diameter of single particles does not depend on temperature. The fact that its value practically coincides with the data of the magneto-granulometric analysis lends support to the validity of the data obtained. According to the results of [12], nanosized aggregates have very little effect on the optical properties of the solution, since the magnetic cores of the particles are separated by a double layer of the stabilizer, and the contributions of these particles to the DR signal do not change during aggregation.

The injection of polymer molecules into a colloidal solution can substantially change the properties of the solution. At high impurity concentrations, it leads to the formation of large aggregates, partial precipitation of colloidal particles, and a multiple decrease in the magnetic susceptibility: with increasing impurity concentration, the solution loses its sedimentation stability. Binary solutions are stable to aggregation only at low impurity concentrations (approx. one percent). In this case, their magnetic susceptibility decreases insignificantly. In the DL tests, only sedimentation-stable binary ferrocolloids were used.

A typical dependence of the signal of the measuring cell, assembled according to the standard scheme [12], on the field strength and concentration of impurity molecules is shown in Fig. 3. The signal from a photodiode, which measures the intensity of laser radiation with a wavelength of 633 nm, passing through a thin layer of a magnetic fluid and a polarizer, is plotted on the

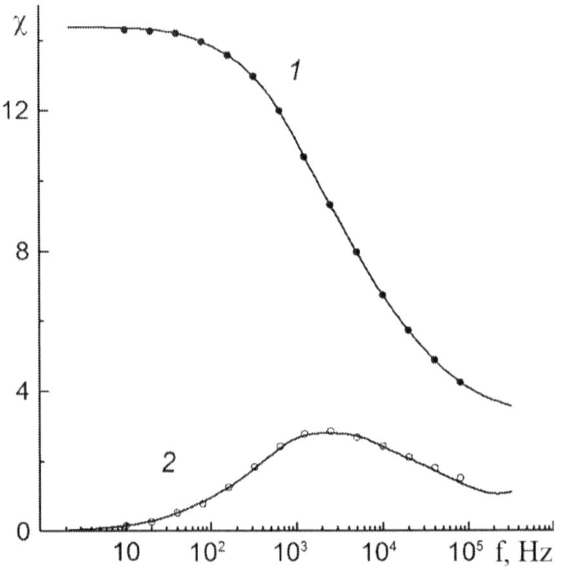

Figure 1. The real (upper) and imaginary (bottom) parts of the dynamic susceptibility of a magnetic fluid versus the frequency of the sounding field at temperature $T = 340$ K

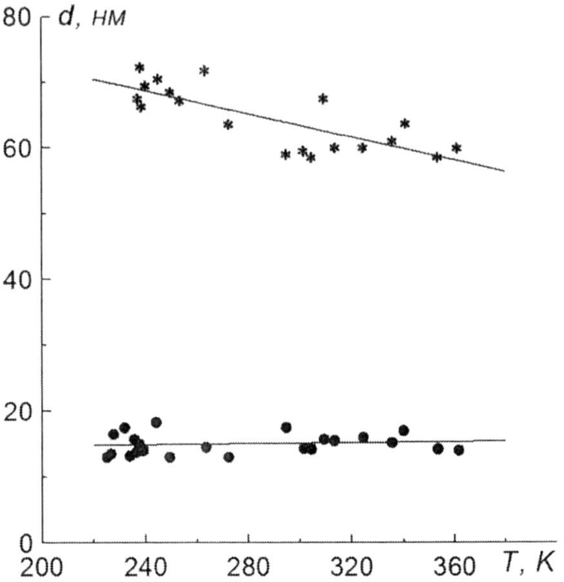

Figure 2. The average diameter of aggregates (top) and single particles in a magnetic fluid as a function of temperature

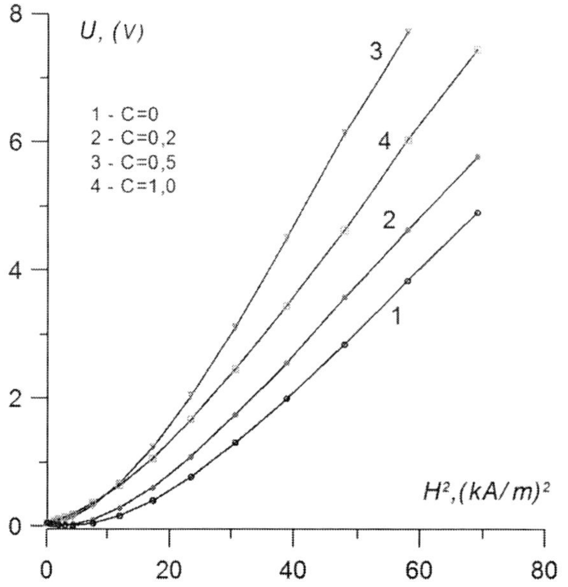

Figure 3. The DR signal as a function of the magnetic field strength for binary ferrocolloids with different concentrations of polymer molecules

ordinate. The constant magnetic field is oriented perpendicular to the light ray. The angle between the plane of polarization of the incident ray and the magnetization vector was $\pi/4$. As expected, the intensity of light passing through the measuring cell increases as the fourth power of magnetic field strength (the difference between the indices of the ordinary and extraordinary rays grows according to a quadratic law). It can be seen from the figure that the injection of polybutadiene molecules into a magnetic fluid can lead to a significant enhancement of DR effect. The maximum (twofold) increase in the DR is observed in the case when the volume concentration of impurity molecules is about 0.5 %.

3. Type II binary ferrocolloids

A binary solution of the second type is synthesized on the basis of a magnetic fluid and impurity nanoparticles of goethite (α−FeOOH). The transverse dimension of the impurity particles (10 − 30 nm) was close to the average diameter of single-domain magnetite particles, and the longitudinal one was an order of magnitude larger. An increase in the DR was due to the orientation of the long axes of the impurity particles along the magnetic field caused by a difference in the "demagnetizing" coefficients along and across the long axis of the impurity particle. The magnetic birefringence was studied as a function of the concentration of the magnetic phase, impurity particles, and magnetic field strength. The main result of the study was the experimental substantiation of a multiple amplification of the DR signal by impurity particles, which was predicted in calculations made in [17]. For the sake of illustration, Fig.4 shows the DR signal for a binary solution and a base magnetic fluid as a function of the magnetic field strength. In weak fields and for a volume fraction of impurity particles of the order of one percent, the DR signal intensity is increased by one or two orders of magnitude. In the fields of about 10 kA/m, the signal gain, associated with the influence of impurity particles, reaches saturation and in the future practically does not change. The total optical anisotropy continues to increase due to the orientation of magnetite particles. A surprising result is that curve 2 in Fig. 4 (in contrast to curve 1 for a magnetic fluid) is close to a linear relationship, and there is

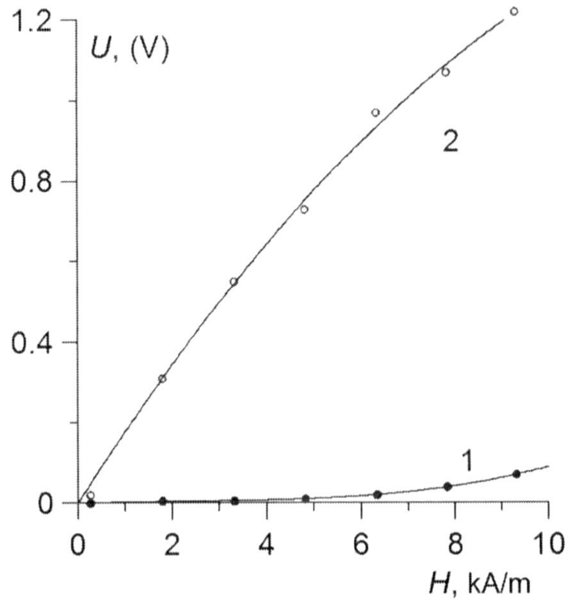

Figure 4. The DR signal in experiments with a binary ferrocolloid of the second type. 1 - magnetic fluid, 2 - solution with α−FeOOH particles ($C = 0.6$ %).

no evidence that in weak fields the signal grows as the fourth power of the field strength. Most likely, the fields of a few kA/m, used in the experiment, can no longer be considered small for impurity particles. In this case, the range of parameters, in which the law $U \sim H^4$ is satisfied, is limited to the field strength of the order of 100 A/m and cannot be resolved in Fig. 4.

Acknowledgments

The work was carried out within the framework of the state budget program N 01201457908 in 2016 − 2018.

References

[1] Skibin Y N, Chekanov V V and Raikher Yu L 1977 Birefringence in a ferromagnetic liquid *J. Exp. Theor. Phys.* **45** 496
[2] Hayes C F and Hwang S R 1977 Observation of magnetically induced polarization in a ferrofluid *Coll. Interf. Sci.* **60** 443
[3] Sholten P C 1980 The origin of magnetic birefringence and dichroism in magnetic fluids *IEEE Trans. Magn.* **16** 221
[4] Taketomi S 1983 Magnetic Fluids Anomalous Pseudo Cotton-Mouton Effects about 107 times Larger than that of Nitrobenzene *Jpn. J. Appl. Phys.* **22** 1137
[5] Bacri J C, Perzynski R, Salin D and Servais J 1987 Magnetic transient birefringence of ferrofluids: particle size determination *J. Physique.* **48** 1385
[6] Raikher Yu L and Skibin Yu N 1988 Dynamic birefringence of light in magnetic liquids *Doklady physics (in Russian)* **33** 746
[7] Blums E, Cebers A and Maiorov M M 1997 *Magnetic fluids* (Berlin: Walter de Gruyter)
[8] Taketomi S, Inaba N, Takahashi H, and Miyajima Y 1990 Experimental and Theoretical Investigations on Agglomeration of Magnetic Colloidal Particles in Magnetic Fluids *J. Phys. Soc. Jap.* **59** 3077
[9] Raikher Y L and Stepanov V I 1995 Dynamic Birefringence in Magnetic Fluids. The Effect of Mechanical and Magnetic Degrees of Freedom of the Particles *Europhys. Lett.* **32** 589
[10] Hasmonay E, Depeyrot J, Sousa M H et al. 1999 Optical properties of nickel ferrite ferrofluids *J. Magn. Magn. Mater.* **201** 195

[11] Payet B, Donatini F, and Noyel G 1999 Longitudinal magneto-optical study of Brown relaxation in ferrofluids: dynamic and transient methods *J. Magn. Magn. Mater.* **201** 207
[12] Buzmakov V M and Pshenichnikov A F 2001 Birefringence in concentrated ferrocolloids *Colloid J.* **63** 275
[13] Pshenichnikov A F, Akekseev A A and Lakhtina E V 2006 Magnetic birefringence in binary ferrocolloid *Bulletin of Perm University. Series: Physics* **35** (In Russian).
[14] Lakhtina E V and Pshenichnikov A F 2006 Dispersion of magnetic susceptibility and the microstructure of magnetic fluid *Colloid J.* **68** 294
[15] Pshenichnikov A F, Lebedev A V, Radionov A V and Efremov D V 2015 Magnetic fluid for operation in strong gradient fields *Colloid J.* **77** 196
[16] Pshenichnikov A F, Lebedev A V, Lakhtina E V and Kuznetsov A A 2017 Effect of centrifugation on dynamic susceptibility of magnetic fluids *J. Magn. Magn. Mater.* **432** 30
[17] Pshenichnikov A F and Sosnin P A 2002 The magneto-optical properties of an ensemble of ellipsoidal dielectric particles in a magnetic fluid *J. Exp. Theor. Phys.* **95** 275

IOP Conf. Series: Journal of Physics: Conf. Series **994** (2018) 012011 doi:10.1088/1742-6596/994/1/012011

Structural Studies of dielectric HDPE+ZrO_2 polymer nanocomposites: filler concentration dependences

A A Nabiyev[1,2], A Kh Islamov[2], A M Maharramov[1], M A Nuriyev[1], R S Ismayilova[1], A S Doroshkevic[2,5], A Pawlukojc[2,4], V A Turchenko[2], Olejniczak A[2,7], M İ Rulev [2,3,6], V Almasan[7] and A I Kuklin[2,3]

[1]ANAS Institute of Radiation Problems, 9 B. Vahabzade str., AZ1143 Baku, Azerbaijan

[2]Joint Institute for Nuclear Research, Joliot-Curie 6, 141980 Dubna, Moscow region, Russia

[3]Moscow Institute of Physics and Technology, MIPT, 141701 Dolgoprudny, Russia

[4]Institute of Nuclear Chemistry and Technology, Dorodna Str16, 03-195 Warsaw, Poland

[5]Donetsk Institute for Physics and Engineering named after O.O.Galkin of the NAS of Ukraine, 03680, Naukiave, 46, Kyiv, Ukraine

[6]European Synchrotron Radiation Facility, F38043 Grenoble, France

[7]Faculty of Chemistry, Nicolaus Copernicus University, ul. Gagarina 7, 87-100 Torun, Poland

[8]National Institute for Research and Development of Isotopic and Molecular Technologies, 67-103 Donat PO5 Box 700, 400293 Cluj-Napoca, Romania

E-mail: asifnebi@jinr.ru

Abstract. Structural properties of HDPE+ZrO_2 polymer nanocomposites thin films of 80-100μm thicknesses were investigated using SANS, XRD, Laser Raman and FTIR spectroscopy. The mass fraction of the filler was 1, 3, 10, and 20%. Results of XRD analysis showed that ZrO_2 powder was crystallized both in monoclinic and in cubic phase under normal conditions. The percentages of monoclinic and cubic phase were found to be 99.8% and 0.2%, respectively. It was found that ZrO_2 nanoparticles did not affect the main crystal and chemical structure of HDPE, but the degree of crystallinity of the polymer decreases with increasing concentration of zirconium oxide. SANS experiments showed that at ambient conditions ZrO_2 nanoparticles mainly distributed like mono-particles in the polymer matrix at all concentrations of filler.The structure of HDPE+ZrO_2 does not changes up to 132°C at 1-3% of filler, excepting changing of the polymer structure at temperatures upper 82°C. At high concentrations of filler 10-20% the aggregation of ZrO_2 nanoparticles occurs, forming domains of 2.5μm. The results of Raman and FTIR spectroscopy did not show additional specific chemical bonds between the filler and the polymer matrix. New peaks formation was not observed. These results suggest that core-shell structure does not exist in the polymer nanocomposite system.

1. Introduction

In recent years a new generation of polymers such as high density polyethylene (HDPE) is being used as a matrix. The base polymers are reinforced with nanoparticle additives in order to improve the properties of polymer nanocomposites. In this case it becomes possible to adjust the structure and properties of the materials in wide range. The adding of inorganic spherical nanoparticles to polymer matrix allows the modification of the properties of polymers. These properties include physical, chemical, thermo mechanical, dielectric, optical and magnetic. Lastly, typical potential applications will be shown with the focus on new and innovative applications, like in energy storage systems [1].
ZrO_2 is one of the interesting oxide materials which is used as nanoparticle filler of the base polymers in order to improve their properties. This oxide material due to its hardness, optical transparency and high refractive index [2, 3] has already found a wide range of applications in different fields. These materials are used as interferometry filters in photonics, as a coating high power laser mirrors and atsetra. Pure ZrO_2 is a polymorphic material existing in three possible cristalline forms: monoclinic, tetragonal and cubic. The monoclinic phase is stable below 1170°C , the tetragonal at the temperature range of 1170-2370°C and the cubic at over 2370°C [4, 5, 6]. High strength and functionality make ZrO_2 filled nanoporous polymeric materials to develop highly relevant and attractive from an economic point of view. The unique stability of the ZrO_2 material to radiation opens up the prospects of creation on their basis of light high-strength and relatively inexpensive materials. This property can provide effective radiation protection of the navigation instruments of the spacecraft, instrument panels special equipment, working in conditions of high radiation background. The aim of this work is a comprehensive study of the structure of ZrO_2 filled nanoporous polymeric materials based on polyamides with the use of unique nuclear-physical methods. These studies will help to obtain new scientific knowledge and to approach the practical implementation of new structural and functional devices based on them [7, 8].
Polymer nanocomposites as a separate material, has a number of promising electrophysical [9], electret [10] and optical properties in applied approaching. These properties make them suitable for using for development for a new sensors, photovoltaic cells, piezomaterials, etc [11]. These types of nanocomposite materials have strong physical properties in comparison with ordinary polymers and leads to find wide application area. Polymer nanocomposites materials, based on a polymer matrix and inorganic ceramic nanoparticle fillers, have drawn considerable attention due to improvements in various properties. These properties include electrical, thermal, optical and other mechanical. The polymer nanocomposites materials are important to the electronic industry for its dielectric properties when used in capacitors. Polymer nanocomposites are considered constructial and industrial materials of the 21st century [12,13,14].
Addition of nanoparticles into polymer matrices remarkably changes thermal mechanical, optical, and electrical properties relative to the host polymer. Among these components are advanced capacitors that greatly require the advanced dielectric materials and film processing techniques. It has been well recognized that every new generation of capacitors have been primarily credited to the innovation and engineering of new dielectric materials [15].
The goal of this work was the investigation of the structure and interactions of HDPE+ZrO_2 in wide range of concentrations and temperatures.

2. Materials and methods

New high density polyethylene matrix samples (HDPE+ZrO_2) were obtained using a crystal zirconium dioxide (ZrO_2) particles of size 20-30 nm (Sky Spring Nanomaterials, Inc. Houston, USA [16]) with spherical morphology by thermal pressing method under pressure of 15 MPa and at a temperature of 165°C and then a sharp cooling in the water-ice system. Studied samples were made as thin film form. The spherical shaped ZrO_2 nano particles (20 ~ 30 nm) were used as a filler. The mass fraction of the nano filler was 1, 3, 10, and 20% in the polymer matrix. The thickness and the diameter of the samples were 80-100 μm and 5 cm, respectively. As a polymer matrix, it has been chosen powdered high-density polyethylene (powder-shaped stamps 20806-024) with an average molecular weight of $95 \cdot 10^3$ g/mole, crystallinity degree of 65%, melting point of 130°C, and density of 958 kg/m^3. Selection of HDPE as a matrix is due to the fact that, it has a unique set of physical and mechanical properties, high wear

properties and impact strength, stability in harsh environments. The choice of HDPE is due to the low cost, adaptability, abundance and good dielectric properties [17, 18].
As a filler it has been used a crystal zirconium dioxide ZrO_2 spherical particles (Sky Spring Nanomaterials, Inc. Houston, USA) with size of 20~30 nm, specific surface area of S=35 m^2/g and density of 5.68 g/cm^3 [16]. Studied samples were made as thin film, the thickness and the diameter were 80-100μm and 5 cm, respectively.

Small-angle neutron scattering (SANS): Small-angle neutron scattering (SANS) measurements were performed at the time-of-flight YuMO spectrometer situated at IBR-2 pulsed reactor, JINR, Dubna, Russia [19, 20]. The experiments were carried out at sample-to-detector distances of 5.28 and 13.04 m, resulting in a Q range of 0.007– 0.4 $Å^{-1}$) in the temperaturerange from 25°C to 132°C. The samples were kept in a special thermal box connected to a Lauda computer-controlled thermostat. The diameter of the sample in the beam was 14 mm. The measured neutron scattering spectra were corrected for the transmission and thickness of the sample, background scattering on the film substrate and on the vanadium reference sample using SAS software [21, 22, 23], yielding a neutron scattering intensity in absolute units of cm^{-1}. Measurements were carried out in the temperature range from 25°C to 132°C.

X-ray diffraction: The definition of phase composition and type of crystal structure has been performed on the conventional X-ray diffractometer EMPYREAN (firm PANalytical) in Cu- Kα radiation with Ni filter in the room temperature (25°C). The operation mode of the device was 40 kV voltage and 30 mA current. The scanning has been done with a step of 0.026° and 2 s per step for 2θ. Analysis of phase composition and type of crystal structure of composite material has been performed using ICDD database.

RAMAN spectroscopy: Raman spectroscopic analyses can be applied for various materials, such as nanoparticles and polymer composites. This method is an important analytical and research tool which provides key information about polymer nanocomposites and material structure. Raman spectra were collected on a Horiba Lab RAM HR evolution spectrometer. Spectra were obtained in the wavelength range 3500-50 cm^{-1}. The instrument has two lasers: a HeNe laser with the emission lines at 633 nm and 325 nm from He-Cd. The HeNe laser was focused on the sample with a 100× objective in the experiment. Filter was not used during the measurements. HORIBA Scientific's Lab Spec 6 software was used for complete instrument control and data processing.

FTIR Spectroscopy: To study structural changes in the samples, was used the technique of IR spectroscopy based on the measurement of transmission spectra of semi-transparent films. Spectra were obtained by the FTIR spectrometer of Shimadzu FT-IR in the wavelength range 4000-750 cm^{-1}. Samples for study were obtained in the form of thin films with a thickness of 80-100 microns size 8x30 mm^2 stage-by-stage pressing at a pressure of $6x10^4$ N/m^2 at room temperature using a special mold of Shimadzu.

3. Experimental

XRD study

X-ray diffraction (XRD) patterns of HDPE+20%ZrO_2, ZrO_2 powder and pure HDPE are shown in Figure 1. Observable diffraction lines of ZrO_2 coincide with monoclinic structure P 21/c1 of baddeleyite described early in reference [24, 33]. As shown in figure HDPE has two diffraction peaks when 2θ is between 20° and 25°. The 2θ with corresponding crystalline lattices for HDPE are 21.59°, 23.93°, 30.08° and 36.29° being composed of orthorhombic crystals and 2θ =19.40° triclinic [32].
HDPE+ZrO_2 nanocomposites and ZrO_2 have identical crystalline lattices, except for 2θ = 21.59°. ZrO_2 does not affect the main crystal structure of HDPE. As a result, ZrO_2 and HDPE acquires monoclinic and orthorhombic phase, respectively.
From X-ray diffraction measurements it is determined that the structure of pure ZrO_2, as well as the structure of HDPE + 20% ZrO_2 nanocomposites has monoclinic symmetry and belonged to the space group of P1 21/c1 (14). The following values have been calculated for the crystal lattice parameters: a=5.149Å, b=5.210Å, c=5.318Å, α=γ=90°, β=99.215°, Z=4, V=140.832$Å^3$. D spacing (lattice period) for the samples ranged between 1.4 – 5 Å. The results of XRD studies showed that the nano filler does not affect the crystal structure of the matrix when it is added to the polymer matrix at 1-3% of

concentrations, whereas at 10-20% structure of polymer became less crystalline, several diffraction peaks from polymer disappear.

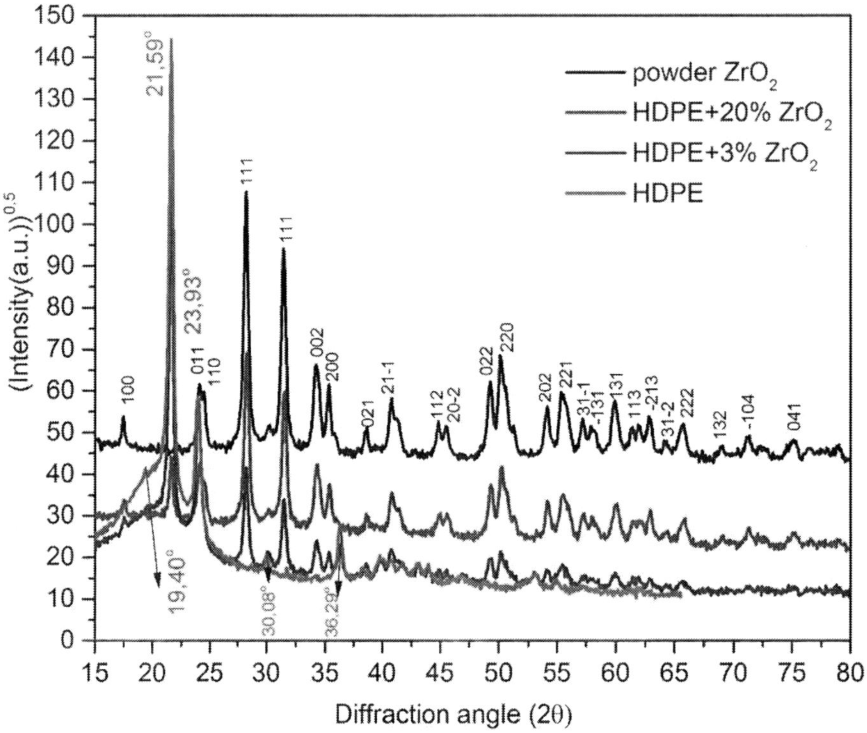

Figure 1. X-ray diffraction (XRD) patterns of pure HDPE (violet), pure ZrO_2 powder (black), HDPE+3%ZrO_2 (blue) and HDPE +20% ZrO_2 (red) polymer nanocomposites

SANS analysis

The small-angle neutron scattering curves of HDPE+ZrO_2 (1, 3, 10 and 20%) and the pair distribution functions P(R)obtained from program GNOM (inset) are shown in Figure 2.

It can be seen that P(r) does not change much as the ZrO_2 volume fraction increases. Total excluded volume V_p (the aggregation volume), maximal length of nanocluster, number of aggregation (N_{agg}) were obtained using the data analysis software ATSAS 2.8.2 software program DAMMİF and Gnom [25, 26, 27]. The results are shown in Table 1. The average distance D between ZrO_2 aggregates in the polymer matrix (1-20%) was obtained from eq.1, using the approximation of a primitive cubic lattice:

$$D^3\varphi = V_p \qquad (1)$$

Where, D is the mean distance between nanoclusters, φ is the volume fraction of the filler and V_p is the average volume of the nanocluster (see tab.1).

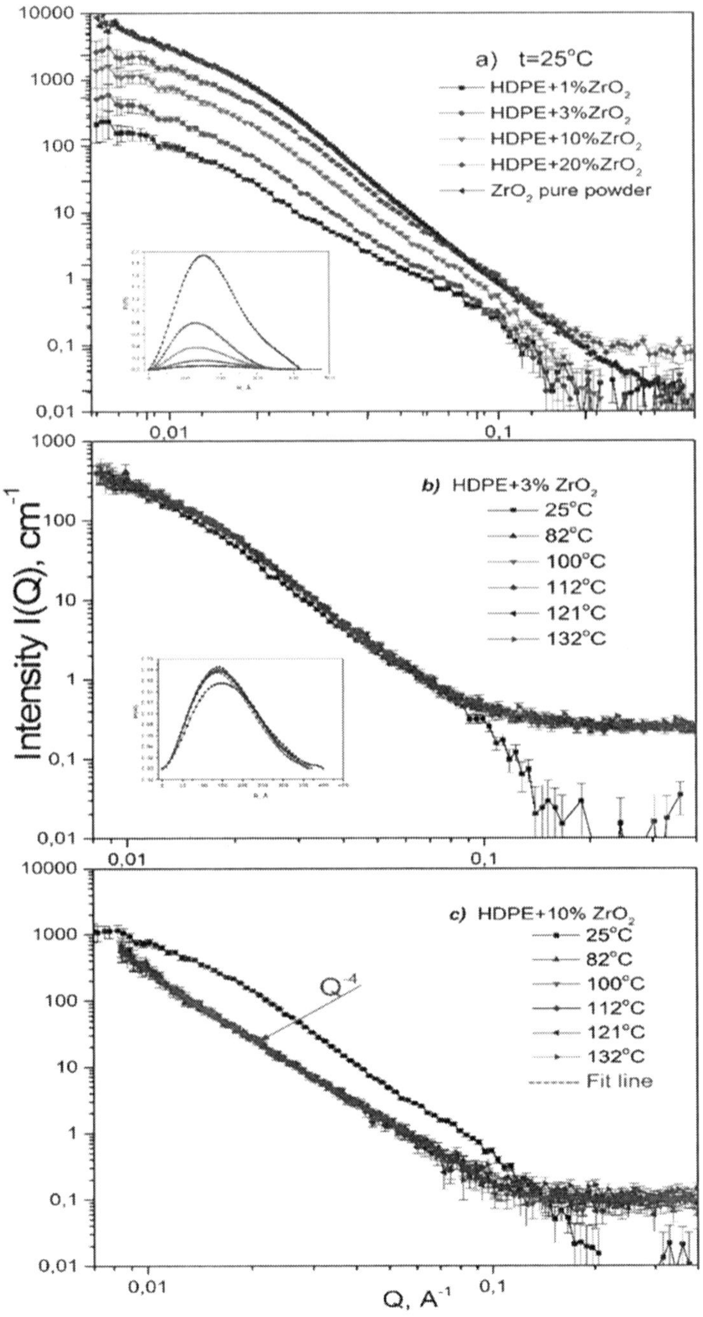

Figure 2. The small-angle neutron scattering curves of HDPE+ZrO_2, the pair distribution functions P(R) obtained from program GNOM (inset): **a)** 1,3,10 and 20% ZrO_2 at 25°C; ***b)*** 3%ZrO_2 at 25 - 132°C; ***c)*** 10%ZrO_2 at 25 -132°C

It is seen that the aggregation practically does not change with increasing of the ZrO_2 filler concentration and is about 1.5 of nanoparticles whereas for pure ZrO_2 powder the aggregation is about 4 nanoparticles. It means that for all concentrations of filler ZrO_2 nanoparticles are distributed in the polymer matrix homogeneously as monomers and separated by a high density polyethelene interlayer.

Table 1. SANS results obtained using the software DAMMİF and Gnom. Total excluded volume V_p, maximal length of nanocluster, the number of aggregation N_{agg} ($N_{agg}=V_p/V_m$, here V_p- total excluded volume, V_m-volume of monomer) in the nanocluster. The volume fraction φ of ZrO_2, D the average distance between nanoclusters for ZrO_2 in nanopowder / polymer matrix composites (1-20%).

HDPE+%ZrO_2	Total excluded volume V_p, nm^3	Maximal length of nanocluster, nm	Number of aggregation, N_{agg}	Volume fraction, φ	D the average distance between nanoclusters, nm
1	$0.876*10^4$	51.6	1.2	0.0018	169.5
3	$1.27*10^4$	61.6	1.7	0.0053	133.8
10	$1.12*10^4$	51.8	1.5	0.018	85.37
20	$1.07*10^4$	55.3	1.5	0.035	67.36
ZrO_2 powder	$2.82*10^4$	85.8	4	---	---

As it is seen from Table 1. at 25°C ZrO_2 nanoparticles are distributed in the polymer matrix homogeneously as monomers and separated by a high density polyethelene interlayer for all concentrations of filler.

At small concentration of (1-3%) $ZrO_2$the scattering behavior of polymer nanocomposite system does not changes with increasing of temperature up to 132°C (see Figure 2 b)), excepting the range of large Q scattering vectors–the region responsible for the scattering from the polymer matrix. The possible reason is that polymer became more amorphous at high temperatures. The changes of polymer structure in the composite system with temperature needs further detailed investigations.

At highconcentrations 10-20% the scattering behaviour changes significally with temperature (see Figure 2. c)). The aggregation of ZrO_2 nanoparticles occurs with melting of polymer matrix.

At 25 °C scattering definitely shows Guinner region, whereas at the temperatures higher 82°C the scattering follows to Porod law:

$$I(Q) = 2\pi\left(\rho_{ZrO_2} - \rho_{HDPE}\right)^2 \cdot Q^{-4} \cdot \frac{S}{V} = 2\pi\left(\rho_{ZrO_2} - \rho_{HDPE}\right)^2 \cdot Q^{-4}\,\frac{3}{R} \qquad (2)$$

Where ρ_{ZrO2} and ρ_{HDPE} are the scattering densities of ZrO_2 and polyethelene matrix. S and V are the surface and volume of the object with smooth surface, for spherical object S/V ratio is 3/R , where R – is the radius of sphere. The spherical estimation of aggregation size from Porod law gives around 2.5μm.

Raman spectroscopy

Figure 3. shows Raman spectra for HDPE+(1-10%) ZrO_2, ZrO_2 powder and pure HDPE. The analysis of the Raman spectra showed that peak intensities depend on the filler contents.

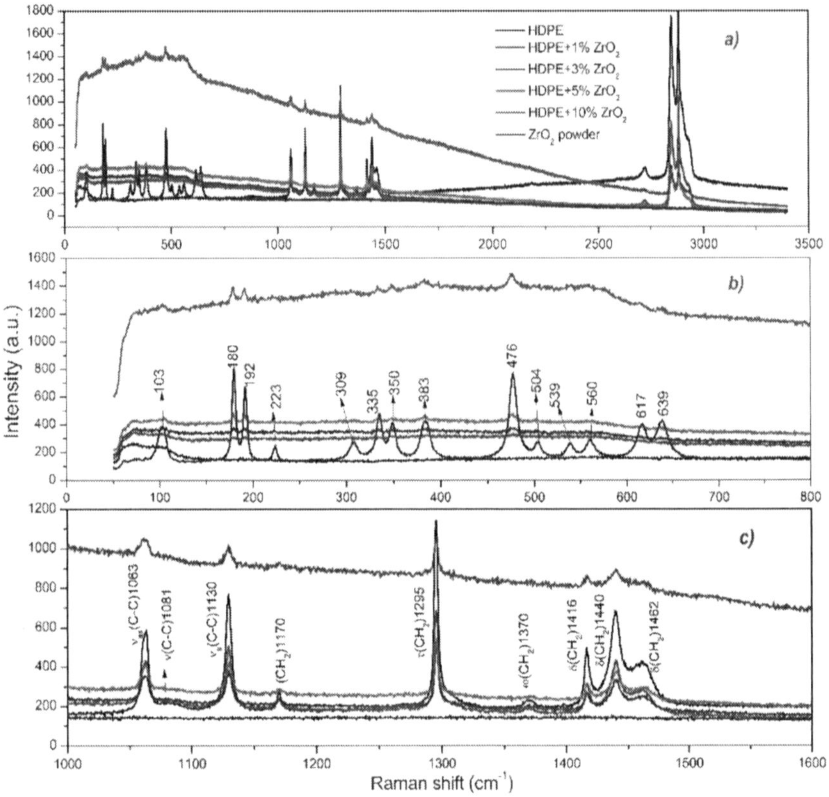

Figure 3. Raman results for HDPE+(1-3-5-10%)ZrO_2, ZrO_2 powder and pure HDPE. a) 50 - 3500 cm^{-1}, b) 0 - 800 cm^{-1} and c) 1000 – 1600 cm^{-1}interval was shown.

It should be noted that amorphous ν stretching C-C bonds are clearly seen for the with 1-3% filler content at 1081 cm^{-1} point, but this is absent at 10% filler contents. The same process is also seen in crystalline+amorphous C-H_2 bonds at 1170 and 1370 cm^{-1} for the with 10% filler content. However, the peaks in the range of 50 to 800 cm^{-1} are assumed to be related to Zr-O- Zr bonds. These peaks are shown at 103, 180, 192, 335, 350, 383, and 476 cm^{-1}. As seen in Fig 3. b in the 50-800 cm^{-1} spectral region at 25°C temperature the observed peaks as external mode 103, 180, 192, 223 cm^{-1}, internal mode 476, 504, 539, 560, 617, 639, 758 cm^{-1} and intermediate spectral region 309, 335, 350 and 383 cm^{-1} are the same as reported in this work [28]. The intensity of the peaks increases depending on the concentration of ZrO_2. This increase manifests for the 10% filler content. Vibrational phase and modes of the Raman spectrum of HDPE+(1-10%)ZrO_2 nanocomposites and pure high density polyethyleneare summirized in the table 2. As shown in Fig 3. c in the 1000-1600 cm^{-1} spectral region the Raman bands are assigned to the crystalline chain at 1063, 1130, 1295 and 1416 cm^{-1} and the amorphous chain at 1081, 1440 and 1462 cm^{-1} points. In the Fig 3. **b** in the 1000-1600 cm^{-1} spectral region it is shown three distinctive peaks between 1400 -1500 cm^{-1} (1416, 1440 and 1462 cm^{-1}). These peaks are assigned to the CH2 bending modes. The band at 1416 cm^{-1} is also useful for estimation of the crystallinity of polyethylene. The 1440 cm^{-1} and 1462 cm^{-1} band are due to the CH2 bending modes of the amorphous trans chains and the melt-like amorphous chains, respectively. The Raman bands at 1295 cm^{-1} and 1370 cm^{-1} are assigned to the CH2 twisting modes of the crystalline and amorphous chains, respectively. These bands are used in the polymer matrix to estimate the crystallinity and molecular orientation. These results are very close to the results of authors [29-30] and confirm their results.

Table 2. Vibrational phase and modes of the Raman spectrum of pure HDPE and HDPE+(1-10%)ZrO_2 nanocomposites. Here is ν_{as}– asymmetric stretching, ν_s – symmetric stretching, τ- twisting, δ- bending, ω- wagging chemical bonds.

Raman shift, cm^{-1}	Mode	Phase
1063	ν_{as}(C-C)	Crystalline amorphous trans chain
1081	ν(C-C)	amorphous
1130	ν_s(C-C)	Crystalline+amorphous trans chain
1170	ρ(CH_2)	Crystalline+amorphous
1295	τ(CH_2)	Crystalline
1370	ω(CH_2)	Crystalline+ amorphous
1416	δ(CH_2)	Crystalline(Orthorhombic)
1440	δ(CH_2)	Amorphous trans (interphase)
1462	δ(CH_2)	Amorphous(Intermediate)

FTIR Spectroscopy

Figure 4. Presents FTIR results for HDPE+ (1-3-5-10-20%) ZrO_2 and pure HDPE in the interval 4000 - 750 cm^{-1} (see Figure 4. a). The bands chosen for HDPE+ZrO_2 polymer nanocomposites analysis were in the regions 700-750, 1400–1500, 2300-2400 and 2800–3000 cm^{-1}. In the 700-750 and 1420-1500 cm^{-1} regions are shown doublet peak at 719.28, 731.07, 1462.07 and 1471.6 cm^{-1} points (see Figure 4. b and c).

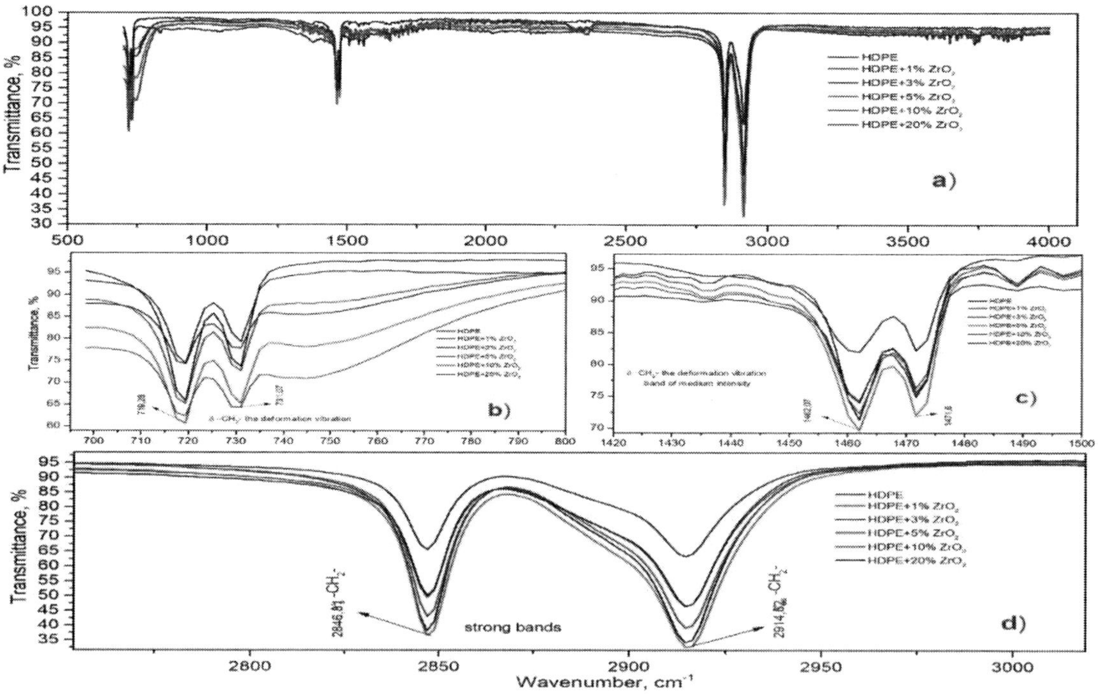

Figure 4. FTIR spectra of HDPE and HDPE+ (1-3-5-10-20%)ZrO_2 polymer nanocomposites

The absorbance peaks are located in a wide range of spectra scale. The first peak at 719.4, 729.5cm^{-1} is C-H rocking deformation vibration, the second peak is for C-H bending deformation at 1461.6 and 1471.4cm^{-1} doublet peak occurred due to the presents of some impurities on the sample. The next peak that is located at 1461 cm^{-1} respectively indicates aromatic C-H bond stretching vibration. The next peaks that are located in the region 2800-2950 cm^{-1} respectively indicate strong CH_2 asymmetric and symmetric stretching vibration (see Figure 4.d). The peaks at 1306.1 cm^{-1} corresponds to twisting deformation, 1367.3 cm^{-1} to wagging deformation vibration [31].
The results of Raman and FTIR studies showed that there are no additional peaks in obtained HDPE+ZrO_2 spectra compared with pure ZrO_2 and HDPE samples. It indicates that there are no specific chemical bonds between the nano filler and the polymer matrix or the core-shell structure is not found in the polymer nanocomposite.

4. Conclusion

The results of XRD studies showed that the nano filler does not affect the crystal structure of the matrix when it is added to the polymer matrix at 1-3% of concentrations. At 10-20% of filler content, the structure of polymer becomes less crystalline as results of some of the diffraction peaks from polymer disappear. Any structural change does not occur. It is found that nanoparticles and HDPE have a monoclinic and orthorhombic structure in the polymer matrix.
SANS experiments showed that at ambient conditions ZrO_2 nanoparticles are mainly distributed as monomers in the polymer matrix and separated by a high-density polyethylene interlayer for all concentrations of filler. The number of aggregation is about 4 for the ZrO_2 nano powder. The structure of ZrO_2 nanoclusters does not change up to 132°C at 1-3% of filler content excepting changing of the polymer structure at temperatures upper 82°C. At high concentrations of filler (10 - 20%) the aggregation of ZrO_2 nanoparticles occurs starting from 82°C, forming ZrO_2 domains of 2.5 μm.
The results of Raman and FTIR spectroscopy showed that there are no additional chemical bonds between the filler and the polymer matrix. New bands formation was not observed. The results showed that core-shell was not found in the polymer nanocomposite.

Acknowledgement
The work was partially supported through the JINR-Romania Cooperation Programme Project of 2017 Order No. 219/55.

References

[1] Thomas Hanemann, and Dorothée Vinga Szabó 2010 *J.Materials* **3** 3468-3517
[2] M R N Soares, C Nico, D Oliveira, M Peres, L Rino, A J S Fernandes, T Monteiro, F M Costa 2012 *Materials Science and Engineering* B **177** 712–716
[3] S Kumar, S Bhunia, A K Ojha 2015 *Physica* E **66**, 74-80
[4] J C Garcia, L M R Scolfaro, A T Lino at. al 2006 *J. Appl. Phys.* **100** 104103-104109
[5] Chevalier J, Gremillard L, Virkar A V at. al 2009*J. Am. Ceram Soc* **92** 1901-1920
[6] Suresh A, Mayo M J, Porter W D at. al 2003 *J. Am. Ceram.* **86** 360-362
[7] Geethalakshmi K, Prabhakaran T and Hemalatha J 2012 *International Journal of Chemical, Molecular, Nuclear, Materials and Metallurgical Engineering* **6 4** 256-259
[8] Shruti Nambiar and John T W Yeow 2012 *ACS Appl. Mater. Interfaces* **4** 11 5717–5726
[9] Tyutnev A P, Saenko V S, Pozhidaev E D at. al 2005 Dielectric properties of polymers in the fields of ionizing radiation Under ed. N S Kostyukova Book **5** M Nauka 453
[10] Gorokhovatskiy Y A, Aniskina L B, Burda VV et. al 2009 *Proceedings of Russia pedag.university named after A.I.Gertsen, Scientific journal* **95** 63-66
[11] Guo N, Sara A Di Benedetto at. al 2010 *J.Chem. Mater.* **22** 4 1567-1578
[12] Najiba Abdullah Al-Hamadani, Noor Dhaief Al-Shwak 2015 *IOSR Journal of Applied Physics*

IOP Conf. Series: Journal of Physics: Conf. Series **994** (2018) 012011 doi:10.1088/1742-6596/994/1/012011

72278-4861
[13] Selyutin G E, Gavrilov Y Y, Voskresenskaya E N et.al 2010 *Chemistry for sustainable development* **18** 375-388
[14] T Tanaka 2005 *IEEE Transactions on Dielectrics and Electrical Insulation* **12** 5 914-928
[15] Daniel Tan, Yang Cao, Enis Tuncer, Patricia Irwin 2013 *Materials Sciences and Applications* **4** 6-15
[16] www.ssnano.com
[17] M M Guliyev, R S Ismayilova, A A Nabiyev 2016 *Journal of Radiation Research* **3** №1 14-23
[18] Guliyev M M, Maharramov A M, Ismayilova R S, Nabiyev A A 2015 *Perspective materials* **7** 17-22
[19] Yu M Ostanevich 1988 *Makromol.Chem.Macromol.Symp.* **15** 91-103
[20] A I Kuklin, A Kh Islamov and V I Gordeliy 2005 *Neutron News* **16** 16
[21] A G Soloviev, T M Solovieva, A V Stadnik, A Kh Islamov, A I Kuklin 2003 *Communication of JINR P10-2003-86*
[22] Kuklin A I, Rogachev A V, Soloviov D V, Ivankov O I, Gordeliy V I 2017 *IOP Conf. Series: Journal of Physics* **848** 012010
[23] Soloviev A G, Solovjeva T M, Ivankov O I, Soloviov D V, Rogachev A V, & Kuklin A I 2017 *IOP Conf. Series: Journal of Physics* **848** 012020
[24] Arashi H, Takeda H, Kudoh Y 1986 *Powder Diffraction* **1** 265 - 275
[25] Svergun D I 1992 *J. Appl. Cryst.* **25** 495-503
[26] D Franke and D I Svergun 2009 *J.Appl.Cryst.* **42** 342-346
[27] D Franke et al. 2017 *J.Appl.Cryst.* **50** 1212–1225
[28] You Ling-Lin, Jiang Guo-Chang, Yang Song-Hua, Ma Jin-Chang, Xu Kuang-Di 2001 *Chin.Phys.Lett* **18** **7** 991-993
[29] Takumitsu Kida, Yusuke Hiejima and Koh-hei Nitta 2015 *IntJ.Exp. Spectroscopic. Tech* 3-6
[30] Saverio Affatato, Enrico Modena, Simone Carmignato, Paola Taddei 2013 *ELSEVIER Wear* **297** 781-790
[31] J.V. Gulmine et al. 2002 *Polymer Testing* **21** 557–563
[32] E R Walter and F P Reding 1956 *J.Polymer.Sci.* **21** 561
[33] Gauna M R, Conconi M S, Gomez S, Suárez G, Aglietti E F, Rendtorff N M 2015 *Journal Ceramics – Silikáty* **59** 4 318-325

Characterization of biogenic ferrihydrite nanoparticles by means of SAXS, SRD and IBA methods

M Balasoiu[1,2*], S Kichanov[1], A Pantelica[2], D Pantelica[2], S Stolyar[3,4], R Iskhakov[4], D Aranghel[1,5], P Ionescu[2], C R Badita[2], S Kurkin[6], O Orelovich[1], S Tiutiunikov[1]

[1] Joint Institute for Nuclear Research, Dubna, 141980 Moscow Region, Russia
[2] Horia Hulubei National Institute for Physics and Nuclear Engineering, Bucharest, Romania
[3] Siberian Federal University, 660041, Krasnoyarsk, Russia
[4] Kirensky Institute of Physics, Siberian Branch of RAS, 660036, Krasnoyarsk, Russia
[5] Extreme Light Intrastructure Nuclear Physics (ELI-NP), Bucharest, Romania
[6] Institute of Synthetic Polymer Materials RAS, Moscow, 117393, Russia

balas@jinr.ru

Abstract. Investigations of biogenic ferrihydrite nanoparticles produced by bacteria *Klebsiella oxytoca* by applying small angle X-ray scattering, synchrotron radiation diffraction and ion beam analysis methods are reviewed. Different experimental data processing methods are used and analyzed.

1. Introduction

A large number of microorganisms produce inorganic nanostructures with controlled size, shape, composition and other properties similar to chemically synthesized materials [1, 2].

In the present work, investigations of biogenic ferrihydrite nanoparticles produced by bacteria *Klebsiella Oxytoca* by applying small angle X-ray scattering (SAXS), synchrotron radiation diffraction (SRD) and ion beam analysis (IBA) methods are reviewed. SAXS enables to measure structural features on length scales between 1 nm up to several hundred of nanometers by analyzing the scattering pattern at very low angles from the direct X-ray beam. By examining X-rays that are scattered at small angles to the primary X-ray beam, it can be obtained information that is directly proportional to the size and shape of nanometer-sized objects [3]. Synchrotron radiation sources (SRS) have several advantages over conventional X-radiation sources, including a wider spectrum and higher luminosity, which allows one to reduce essentially the time of conventional X-ray experiments. The IBA techniques with multi-elemental capabilities, high sensitivity and low detection limits offer perfect methods for elemental characterization of materials [4]. In particular, the simultaneous application of particle-induced X-ray emission (PIXE), particle-induced gamma-emission (PIGE) and Rutherford backscattering spectrometry (RBS) with proton beam can provide quantitative elemental information on a broad range of chemical elements.

2. Experimental

2.1. Biogenic ferrihydrite samples

The investigated samples are ferrihydrite nanoparticles produced by bacteria *Klebsiella Oxytoca* in the course of biomineralization of iron salt solutions from natural environment [5-7]. The bacterial biomass isolated from the sapropel obtained from the Lake Borovoe (Krasnoyarsk region) was grown under microaerophilic conditions on a Lovley medium of composition (g/L) NaHCO3, 2.5; CaCl2·H2O, 0.1; KCl, 0.1; NH4Cl, 1.5; and NaH2PO4·H2O, 0.6. The ferric citrate concentration was varied from 0.2 to 5 g/L, the yeast extract concentration was 0.05 g/L, and the benzoic acid concentration was varied from 0.2 to 0.5 g/L.

As a result of variation of the growth conditions for the microorganisms (growth period, light exposition, potassium citrate – ferric citrate rate, etc.), bacterium *Klebsiella oxytoca* creates, as was established in previous works [6-11] different types of ferrihydrite nanoparticles. Two types of ferrihydrite nanoparticles (Fe12 and Fe34), separated from a bacterial biomass grown during 8 and 21 days, respectively, were accurately identified by means of Mossbauer spectroscopy [7] and static magnetic measurements analysis [8].

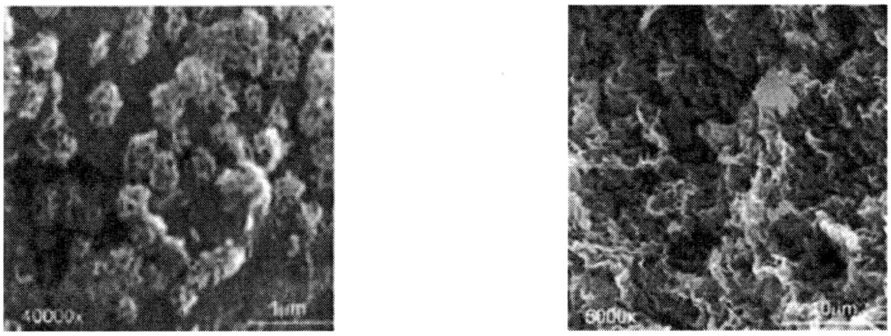

(a) (b)

Fig.1 SEM images of two dry samples containing ferrihydrite nanoparticles obtained from a bacterial biomass grown during 8 and 21 days, respectively; sample Fe12 (a) and sample Fe34 (b).

Earlier studies [1, 11] of the morphology and structure investigation of these two ferrihydrite samples produced by *Klebsiella Oxytoca* by means of a number of procedures at different length scales: optical microscopy; scanning electron microscopy (Figs.1 a, b), transmission electron microscopy and small angle X-ray scattering have given the results presented in [1-2, 12].

Spectroscopic studies (absorption, fluorescence emission, and FT-IR) [12-14] confirmed the presence of polysaccharides but also of some traces of proteins.

The preliminary determination of the elemental composition of a ferrihydrite sample grown during 2 weeks was accomplished using X-ray fluorescence analysis (XRF) (see Table 1) [12].

Table 1 Elemental contents determined by XRF in the Fe12 biogenic ferrihydrite sample, in percent d.w. (dry weight) [12]

Element	Fe12
O	60.53
C	16.51
P	8.45
Fe	9.20
Ca	2.93
K	1.68
Cl	0.33
Other elements	0.38

IOP Conf. Series: Journal of Physics: Conf. Series **994** (2018) 012012 doi:10.1088/1742-6596/994/1/012012

In order to characterize the samples with improved analytical sensitivity, also for an analytical quality control, a combined use of different techniques is often required. Within this context, a combined use of PIXE, PIGE, and RBS with 3 MeV proton beam was simultaneously accomplished.

2.2. Small angle X-ray scattering

Small angle X-ray scattering data were obtained at Brucker Nanostar SAXS spectrometer at the Institute of Synthetic Polymer Materials RAS, Moscow. The X-ray scattering intensity is experimentally determined as a function of the scattering vector Q whose modulus is given by $Q = (4\pi/\lambda)\sin\theta/2$, where λ is the X-ray wavelength, and θ is the scattering angle between the directions of the scattered and transmitted beams. The experimental setup used covered the Q range of $0.01 \div 0.11$ Å^{-1}.

2.3. Diffraction experiment at SRS

Synchrotron radiation sources have several advantages over conventional X-radiation sources, including a wider spectrum and higher luminosity. This is the reason why synchrotron radiation has been increasingly used in the world to solve a wide spectrum of fundamental and applied problems in physics, chemistry, materials science, biology, medicine and earth sciences.
The Kurchatov Synchrotron Radiation Source (KSRS) in Moscow is the first Russian source put into operation for investigations at X-ray energies up to 40 keV. One of the stations that have been created at the source is the experimental station "Mediana" [15], which hosts facilities for powder diffractometry, tomography and refraction introscopy. The equipment layout at the "Mediana" station is shown in Fig.2.

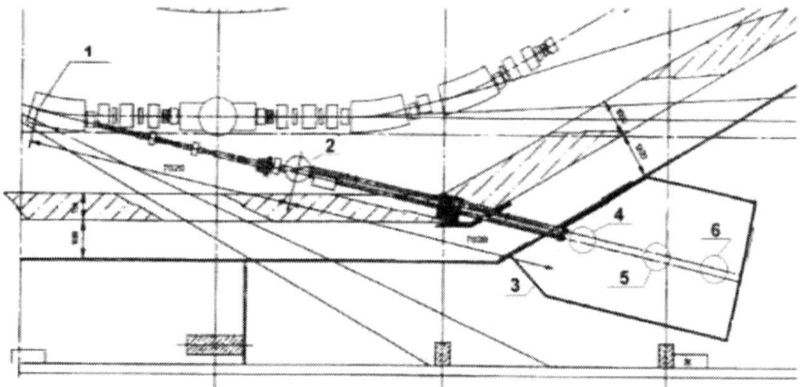

Fig. 2 Layout of the "Mediana" station: 1, radiation source; 2, monochromators; 3, experimental hutch; 4, position of the diffractometer; 5, position of the tomography unit; 6, position of the introscopy unit [15].

One of the topical problems to be solved with the help of synchrotron radiation is the investigation of the structure and phase diagrams of condensed matter. High intensity of the synchrotron radiation beam allows carrying out experiments with small volumes of the investigated matter.

2.4. Ion beam analysis

The 3 MV Tandetron accelerator installed in 2012 at Tandem Accelerators Department (DAT) of the Horia Hulubei National Institute for Physics and Nuclear Engineering (IFIN-HH) is mainly dedicated to applied physics experiments [16]. The machine is fully equipped for ion beam analysis and ion implantation experiments (Fig. 3).

The first beam-line has specific detectors and electronic systems to perform PIXE, PIGE, RBS, and elastic recoil detection analysis (ERDA) (Fig.4).

IBA provides an excellent way to investigate atomic compositions and concentrations in various materials. IBA techniques are characterized by high precision and reliability. In most cases, before the analysis, the samples do not need any chemical or physical preparation, direct rapid analysis is available under vacuum or outside the reaction chamber, in air or in the atmosphere of He to improve the accuracy of the results. The main advantage of this method is its non-invasive profile, which makes it excellent for biological and brittle specimens.

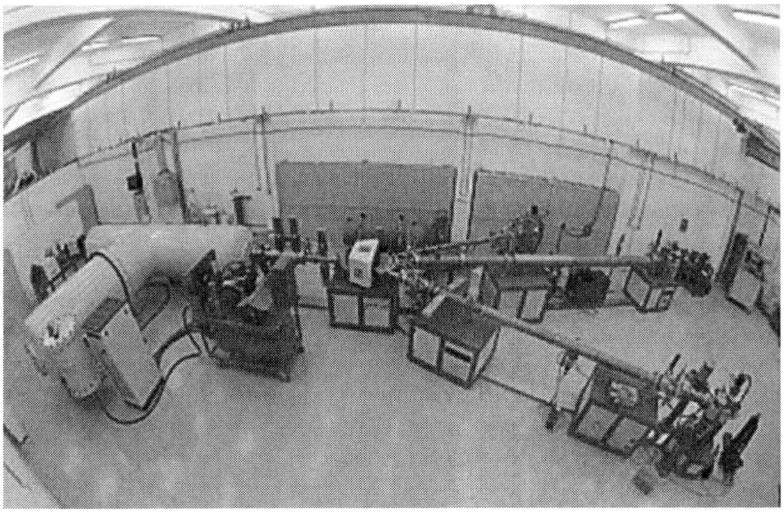

Fig.3 The 3 MV Tandetron accelerator [16] at Tandem Accelerators Department of IFIN-HH [15].

Fig.4 The first beam-line of 3 MV Tandetron accelerator equipped to perform ion beam analysis, e.g. PIXE, PIGE, RBS, and ERDA [16].

3. Results and discussion

3.1. Small angle X-ray scattering

The experimental curve obtained from Sample Fe34 reveals in the q range of 0.01 ÷ 0.11 $Å^{-1}$ a power-law behavior:

$$I(q) \approx q^{-\alpha},$$

with the exponent $\alpha = 3.18 \pm 0.01$. This means that the system has a fractal structure. When $3 < \alpha < 4$, the scattering objects are considered to be surface fractals and the fractal dimension is given by formula $D_S = 6 - \alpha$ [17, 18].

Consequently, for Fe34 sample, the system structure is characterized by a fractal dimension of $D_S = 2.82 \pm 0.01$, which is specific to highly branched surface fractals [9].

For sample Fe12, the experimental curve presents 2 regions with a power-law behavior: 0.01 ÷ 0.022 $Å^{-1}$ and 0.08 ÷ 0.18 $Å^{-1}$. In these cases, the exponent α has the values (i) $\alpha = 2.08 \pm 0.05$ and (ii) $\alpha = 2.86 \pm 0.02$, respectively. For these values, α gives the mass fractal dimension $D_M = \alpha$, characterizing randomly distributed objects as lamellae or platelets (i) [17] and low-dimensional clustering aggregates (ii) [19].

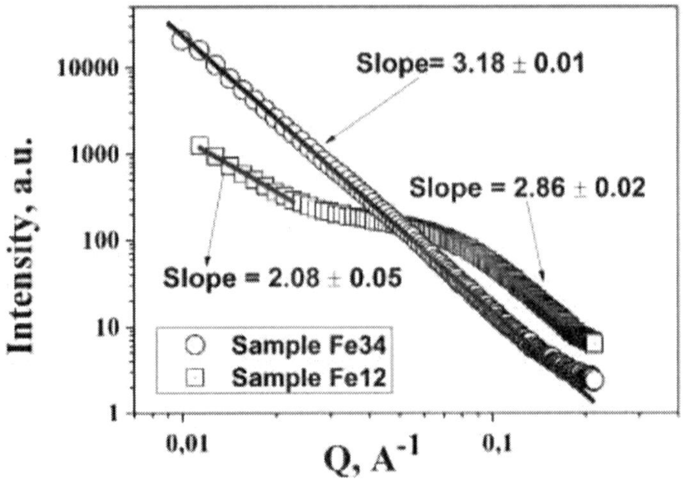

Fig.5 Small angle X-ray scattering experimental curves from dry samples Fe12 (square symbol) and Fe34 (circle symbol) and the power-law fits of the experimental data.

3.2. Diffraction experiment at SRS

The experimental and processed diffraction spectra of biogenic ferrihydrite nanoparticles samples Fe12 and Fe34 obtained at "Mediana" station of the KSRS are given in Fig.6.

The diffraction peaks from the sample are located at scattering angles of $2\theta = 5.4^0$; 8.1^0; 16^0. Besides this, around the scattering angles of $2\theta = 10^0$-13^0 is observed a diffusive halo.

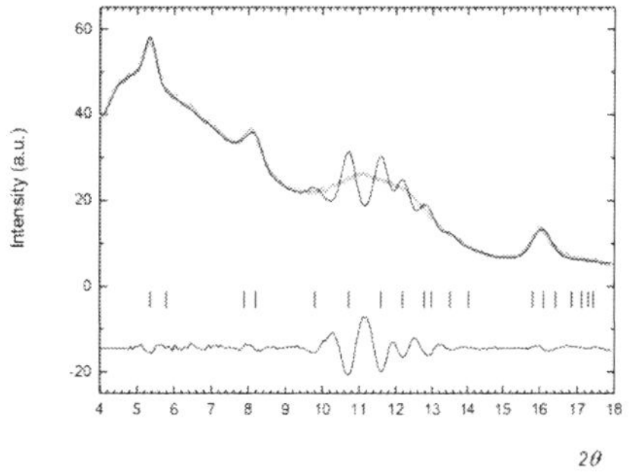

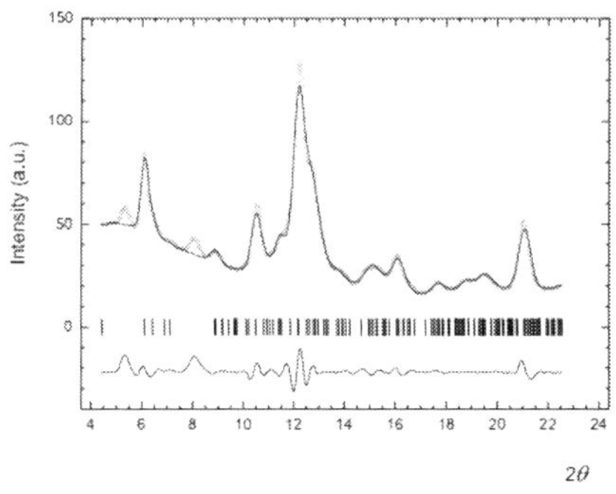

a b

Fig.6 The processing of diffraction spectra sections of: (a) sample Fe12 (using a tetragonal cell model) and (b) of sample Fe34 (using a monoclinic cell) [20].

The processing of a spectrum section of sample Fe12 was accomplished using a tetragonal cell model (a=b=5.81Å, c=6.28 Å). In this case, the peaks located at the scattering angle $2\theta = 5.4^0$; 8.1^0; 16^0 are reflections of the interplanar distances with Miller indices (001) (110) and (003), respectively (see Fig.6 a). In the case of the diffraction spectrum of sample Fe34, a possible description of the structure is the model using a monoclinic cell (a = 3.79, b = 11.04, c = 10.54, β =95.26) (see Fig.6 b). In the diffraction pattern of this sample, one should note the presence of the reflections observed in the spectrum of sample Fe12 at scattering angles $2\theta = 5.4^0$; 8.1^0; 16^0 [20].

3.3. *Ion beam analysis*

PIXE, PIGE, and RBS complementary ion beam techniques were simultaneously used to provide information on elemental content of biogenic ferrihydrite nanoparticles produced by *Klebsiella Oxytoca* [20].

A total 14 elements were determined by PIXE in the examined samples, as follows (concentration levels in descending order, in parentheses): Fe, P, Cl, Ca, and K (tens of percent to percent); Hg, Pb, Al, Si, Rb, and S (tenths of percent); Rb, S, Zn, Cr, and Mn (hundreds of ppm/ part per million).

PIGE results were obtained for Na, Mg, P, Cl, and Fe (5 elements) in Fe12 and Fe34 samples. Major elements Fe, P, and Cl determined by PIGE were found to be at similar levels with those obtained by PIXE (Table 2). In addition, Na and Mg in Fe12 and Fe34 samples were measured by PIGE at levels of (0.80-0.23) % and (1.82-1.10) %, respectively.

RBS results for P, Cl, K, Ca, and Fe are in agreement with those determined by PIXE and PIGE.

By comparing Fe12 and Fe34 samples, higher contents of Fe, P, Si, Hg, Pb, and Rb in Fe34 (ratios between 1.2 and 3), and higher contents of Ca, K, Mg, Na, and Cl in Fe12 (ratios between 1.2 and 10) were found. The elements Cr, and S in Fe34, and Zn and Mn in Fe12 were also determined.

Concerning the light elements determined by RBS, similar mass fractions for O in both samples, 1.5 times higher content for C in Fe34, as well as 1.2 and 2.4 times higher contents for H and N in Fe12, respectively, were observed.

The discrepancy of the elemental contents between IBA results and earlier results obtained by XRD [12] can be explained first by the fact that samples have been extracted differently from the biomass as previously and secondly by the specificity of the investigation method. The cleaning method was improved in order to succeed to eliminate as much as possible the bacterial residues.

Table 2. Elemental contents (percent d.w.) determined by PIXE, PIGE and RBS in Fe12 and Fe34 biogenic ferrihydrite samples grown during 8 and 21 days, respectively.

Element/ Method	Fe12			Fe34		
	PIXE	PIGE	RBS	PIXE	PIGE	RBS
Na	-	0.804 ± 0.040	-	-	0.234 ± 0.012	-
Mg	-	1.82 ± 0.60	-	-	1.10 ± 0.64	-
H	-	-	3.61	-	-	3.08
C	-	-	8.12	-	-	11.9
N	-	-	5.07	-	-	2.14
O	-	-	35.9	-	-	36.7
Al	0.327 ± 0.007	-	-	0.356 ± 0.009	-	-
Si	0.184 ± 0.008	-	-	0.233 ± 0.013	-	-
P	12.3 ± 0.1	10.5 ± 0.5	6.44	15.1 ± 0.1	11.1 ± 0.6	11.2
S	< 0.02	-	-	0.16 ± 0.02	-	-
Cl	11.9 ± 0.1	7.70 ± 1.00	6.86	1.19 ± 0.01	0.89 ± 0.31	0.92
K	6.44 ± 0.02	-	4.10	4.11 ± 0.01	-	3.08
Ca	7.98 ± 0.03	-	5.29	6.67 ± 0.02	-	5.06
Cr	< 0.01	-	-	0.027 ± 0.01	-	-
Mn	0.094 ± 0.003	-	-	< 0.01	-	-
Fe	27.0 ± 0.1	25.9 ± 1.8	24.6	32.7 ± 0.1	30.9 ± 1.5	25.9
Zn	0.039 ± 0.010	-	-	< 0.03	-	-
Rb	0.082 ± 0.016	-	-	0.247 ± 0.030	-	-
Hg	0.328 ± 0.036	-	-	0.684 ± 0.060	-	-
Pb	0.197 ± 0.035	-	-	0.494 ± 0.064	-	-

4. Conclusions

Investigations of biogenic ferrihydrite nanoparticles produced by bacteria *Klebsiella Oxytoca* by applying structural methods as small-angle X-ray and diffraction of synchrotron radiation, as well as ion beam analysis methods as particle-induced X-ray emission (PIXE), particle induced gamma-emission (PIGE) and Rutherford backscattering (RBS) with 3 MeV proton beam were accomplished and elemental composition of samples was determined.

Fe12 and Fe34 samples, separated from a bacterial biomass grown during 8 and 21 days, respectively, showed different crystal structure, i.e. for Fe12, a tetragonal cell model and for Fe34, a monoclinic cell, respectively, give best fits to the their diffraction spectra. At nanoscale level, small-angle X-ray scattering revealed fractal type structural features, i.e. for Fe12, a mass fractal type and for Fe34, a surface fractal type, respectively were obtained.

A total 20 elements were quantitatively determined by ion beam analysis techniques in the investigated samples. From them, 14 elements were determined by PIXE, 5 elements by PIGE and 9 elements by RBS. Common elements measured by different techniques: K and Ca by PIXE and PIGE; P, Cl and Fe by PIXE, PIGE and RBS.

Similar mass fractions for O and Al in both samples, higher contents of Fe, P, Si, C, Hg, Pb, and Rb in Fe34 (ratios between 1.2 and 3), and higher contents of H, Ca, K, Mg, N, Na, and Cl in Fe12 (ratios between 1.2 and 10) were noted. In addition, Cr and S in Fe34, and Zn and Mn in Fe12 were put in evidence.

Acknowledgements
The work was accomplished in the frame of IFIN-HH User Beam schedule at the 3 MV Tandetron accelerator and partially was supported by the Joint Institute for Nuclear Research (JINR) Dubna

Theme No. 04-4-1121-2015/2017, RO-JINR Projects Nos. 95/15.02.2016 and 96/15.02.2016, items 77, 82; Nos. 219/10.04.2017 and 220/10.04.2017 item 38; RFBR and Krasnoyarsk region Project No. 17-43-240527. Support by the Special Program for Siberian Federal University of the Ministry of Education and Science of the Russian Federation is acknowledged.

References

[1] M. Sastry, A. Ahmad, M.I. Khan, R. Kumar, Current Science 85 (2), 162 (2003).
[2] M. Balasoiu, A.I. Kuklin, G.M. Arzumanyan, T.N. Murugova, S.V. Stolyar, R.S. Iskhakov, L.A. Ishchenko, Yu.L. Raikher, Biogenic nanoparticles produced by bacteria Klebsiella oxytoca: structure investigations. In: Modern Trends in Nanoscience, Ed. Maria Balasoiu and Grigory M. Arzumanyan, The Publishing House of the Romanian Academy (Editura Academiei Române), Bucharest, 2013, pp. 179-196.
[3] A. Guinier and G. Fournet, Small-Angle Scattering of X-Rays, John Wiley & Sons, New York, Chapman & Hall, Ltd, London (1995)
[4] J.R. Tesmer, M. Nastasi, Handbook of Modern Ion Beam Materials Analysis. Materials Research Society, Pittsburg, Pennsylvania, USA (1995).
[5] S. V. Stolyar, O.A. Bayukov, D.A. Balaev, R.S. Iskhakov, L.A. Ishchenko, V.P. Ladygina, R.N. Yaroslavtsev, Journal of Optoelectronics and Advanced Materials 17 (7-8), 968-972 (2015).
[6] S. V. Stolyar, O. A. Bayukov, Yu.L. Gurevich, E.A. Denisova, R.S. Iskhakov, V.P. Ladygina, A.P. Puzyr', P.P. Pustoshilov, M.A. Bitekhtina, Inorganic Materials 42, 763 (2006).
[7] S. V. Stolyar, O. A. Bayukov, Yu. L. Gurevich, V. P. Ladygina, R.S. Iskhakov, P.P. Pustoshilov, Inorganic Materials 43, 638 (2007).
[8] Yu.L. Raikher, V.I. Stepanov, S.V. Stolyar, V.P. Ladygina, D.A. Balaev, L.A. Ishchenko, M. Balasoiu, Phys. of Solid State 52 (2) 277 (2010).
[9] M. Balasoiu, S.V. Stolyar, R.S. Iskhakov, Yu.L. Raikher, A.I. Kuklin, O.L. Orelovich, Yu.S. Kovalev, T.S. Kurkin, G.M. Arzumanian, Romanian Journal of Physics 55 (7-8), 782-789 (2010).
[10] M. Balasoiu, L.A. Ishchenko, S.V. Stolyar, R.S. Iskhakov, Yu.L. Raikher, A.I. Kuklin, D.V. Soloviov, T.S. Kurkin, D. Aranghel, G.M. Arzumanian, Optoelectronics and Advanced Materials – Rapid Communications 4 (12), 2136-2139 (2010).
[11] L.A. Ishchenko, S.V. Stolyar, V.P. Ladygina, Yu.L. Raikher, M. Balasoiu, O.A. Bayukov, R.S. Iskhakov, E.V. Inzhevatkin, Physics Procedia 9, 279-282 (2010).
[12] S.V. Stolyar, O. A. Bayukov, V.P. Ladygina, R.S. Iskhakov, L.A. Ishchenko, V.Yu. Yakovchuk, K.G. Dobretsov, A.I. Pozdnyakov, O.E. Piksina, Physics of the Solid State 53 (1), 100–104 (2011).
[13] L. Anghel, M. Balasoiu, L.A. Ishchenko, S.V. Stolyar, T.S. Kurkin, A.V. Rogachev, A.I. Kuklin, Y.S. Kovalev, Y.L. Raikher, R.S. Iskhakov, G. Duca, Journal of Physics: Conference Series 351, 012005, (2012).
[14] C.G. Chilom, D.M. Găzdaru, M. Bălăsoiu, M. Bacalum, S.V. Stolyar, A.I. Popescu, Romanian Journal of Physics 62, 701(13) (2017).
[15] V.L. Aksenov , V.P. Glazkov, S.E. Kichanov, D.K. Pogoreliy, K.M. Podurets, V.A. Somenkov, B.N. Savenko, E.V. Yakovenko, Nuclear Instruments and Methods in Physics Research A 575, 266–268 (2007).
[16] https://tandem.nipne.ro/accelerators.php
[17] G. Beaucage, J. Appl. Cryst. 29, 134-146 (1996).
[18] A.Yu. Cherny, E.M. Anitas, A.I. Kuklin, M. Balasoiu and V. A. Osipov, J. Appl. Cryst. 43, 790–797 (2010).
[19] A.Yu. Cherny, E. M. Anitas, V.A. Osipov and A.I. Kuklin, J. Appl. Cryst. 47, 198-206 (2014).
[20] S. Kichanov, A. Pantelica, D. Pantelica, S. Stolyar, R. Iskhakov, D. Aranghel, P. Ionescu, R. Vladoiu, M. Balasoiu, Romanian Reports in Physics, 70 (2018) (in press).

IOP Conf. Series: Journal of Physics: Conf. Series **994** (2018) 012013 doi:10.1088/1742-6596/994/1/012013

Density functional theory calculations of the water interactions with ZrO_2 nanoparticles Y_2O_3 doped

Mekhrdod Subhoni[1,*], Kholmirzo Kholmurodov[1,2], Aleksandr Doroshkevich[2], Elmar Asgerov[2,4], Tomoyuki Yamamoto[5], Andrei Lyubchyk[6], Valer Almasan[7], Afag Madadzada[2,4]

[1]S.U. Umarov Physical-Technical Institute, Academy of Sciences, Republic of Tajikistan.
[2]Joint Institute for Nuclear Research, 141980, Dubna, Moscow Region, Russian Federation.
[3]Dubna State University, 141980, Dubna, Moscow Region, Russian Federation.
[4]National Nuclear Research Center CJSC, AZ1073, Baku, Azerbaijan.
[5]Waseda University, 169-0051, Shinjuku, Tokyo, Japan.
[6]i3N/CENIMAT, Department of Materials Science, Faculty of Science and Technology, New University of Lisbon and CEMOP/UNINOVA, Campus de Caparica, 2829-516 Caparica, Portugal.
[7]National Institute for Research and Development of Isotopic and Molecular Technologies Cluj-Napoca, Romania 400293 Cluj-Napoca România.

[*]E-mail: mirzo@jinr.ru

Abstract. Development of a new electricity generation techniques is one of the most relevant tasks, especially nowadays under conditions of extreme growth in energy consumption. The exothermic heterogeneous electrochemical energy conversion to the electric energy through interaction of the ZrO_2 based nanopowder system with atmospheric moisture is one of the ways of electric energy obtaining. The questions of conversion into the electric form of the energy of water molecules adsorption in 3 mol% Y_2O_3 doped ZrO_2 nanopowder systems were investigated using the density functional theory calculations. The density functional theory calculations has been realized as in the Kohn-Sham formulation, where the exchange-correlation potential is approximated by a functional of the electronic density. The electronic density, total energy and band structure calculations are carried out using the all-electron, full potential, linear augmented plane wave method of the electronic density and related approximations, i.e. the local density, the generalized gradient and their hybrid approximations.

*Mekhrdod Subhoni-734063, Republic of Tajikistan, Dushanbe, st. Ayni 299/1, Physical-Technical Institute -Technical Institute AS RT, E-mail: mehrdodq@gmail.com

1. Introduction

The water is well known conventional source of electrical energy used by mankind. In liquid state, it is traditionally used for the electric energy conversion by an external thermostat of hydropower plants. However, direct conversion of water in gas state, (atmospheric moisture) to electric energy, remained unrealizable for centuries as a top scientific and practical aim. With the development of nanotechnology, the implementation of these ideas becomes potentially possible In this regard, a nanopowder system based on zirconia dioxide (ZrO_2) is the best candidate for the direct conversion of the energy of water adsorption into electric energy [1]. It is well known [1-2] that the ZrO_2 surface is an extremely nonequilibrium thermodynamically, the surface of the ZrO_2 based nano-dispersed oxide systems is exist in a state of dynamic charge and adsorption equilibration. I.e., the change of quantity of adsorbates is accompanied by a changing of the total charge in the system. In the case of nanoscale powder of zirconia dioxide, on its surface are preferably water is absorbed, and certain stages of a matter exchange between the system and external environment have an exothermic behaviour. Hence, in a cyclic mode, the nanopowder system based on ZrO_2 can convert the chemical energy of water molecules adsorption to an electric form [1-4].

As is known, traditional photoelectric converters use light energy to localize the electron-hole pair in the material of the so-called adsorber (Figure 1). Then the electron-hole pair is converted into a pair of free charge carriers by the heterojunction field.

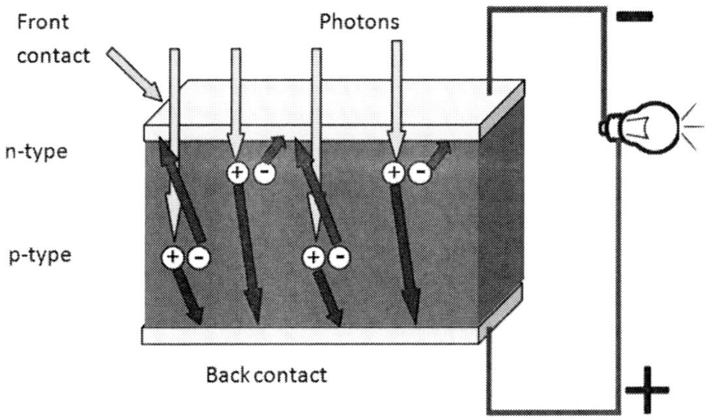

Figure 1. Schematic diagram of the operation of a traditional solar photocell (FEP).

To obtain such a heterojunction, some amount of impurity, with a valence greater than the valence of the base material (intrinsic semiconductor), for example, phosphorus, is introduced into the top layer of silicon. As a result, a layer is formed with an excess of electrons, that is, a negative charge. At the same time, the lower layer is doped with another valence impurity with a valence less than the valence of the intrinsic semiconductor, for example, by boron, which leads to a decrease in the number of electrons, thus creating a positive charge. As a result, an electric field is formed at the junction of these semiconductors, which provides separation and transport of free charge carriers to the electrodes.

The method of obtaining the energy considered in this paper is the phenomenon of injection of hot electrons (the mechanism of transfer of electric energy to the semiconductor crystal, which is realized during the course of a heterogeneous catalytic chemical reaction on its metalized surface) [5] (Figure 2). With a nanometer film thickness of <20 nm, so-called "hot electrons" with an energy of 1-3 eV ballistically, i.e. without loss of energy, reach the Schottky barrier and overcome due to its kinetic energy and, entering the semiconductor, form an electric current in the external circuit. Thus, in this case the appearance of free charge carriers is a consequence not of localization from the lattice, but of tunneling through the heterophase boundary to the outer space.

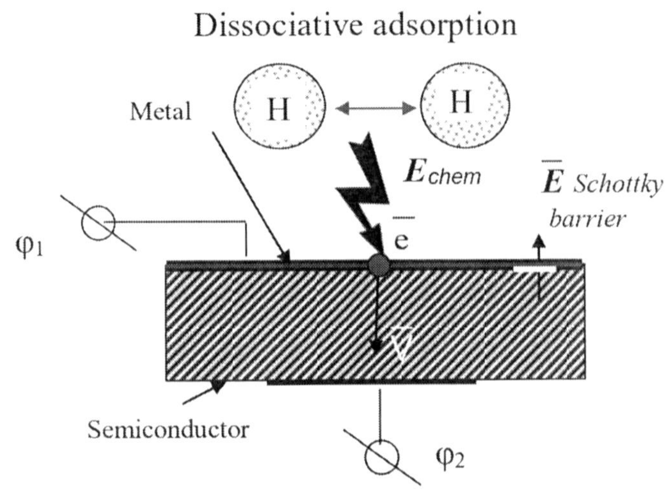

Figure 2. Scheme for the realization of the mechanism of ballistic electron transport on a metal-semiconductor transition during chemical adsorption.

The phenomenon of energy conversion of heterogeneous exothermic chemical reactions by analogy with photo-EMF [3] was called chemo-EMF (the prefix of chemo-, like photo-, emphasizes the nonthermic origin of the phenomenon) [6-9]. This way of converting chemical energy into an electrical form is good because the transformation from the chemical form of energy into electrical energy occurs directly, bypassing the intermediate stages.

There are other ways of obtaining electrical energy from moisture. For example, attempts have been made to collect a charge from jumping droplets of water mist by means of electrostatic capture or causing mechanical vibrations of piezoelectric plates [10-11]. These methods of obtaining energy assume the presence of intermediate energy forms and will have a lower efficiency than in the case of direct energy conversion. However, until now it is not clear in which act of a physic-chemical process a free electron is formed. It can be assumed that when a surface of a nanoparticle interacts with a water molecule, a process similar to that realized in redox processes takes place [12]. This paper is based on the results of the DFT calculations of the structure and energy state of the surface layer of nanoparticles of a solid solution based on ZrO_2 [13], The main tasks of work are:

1. To calculate the energy characteristics of the system, to predict the most probable path of realization of oxidation-reduction processes on the surface of nanoparticles of the investigated disperse system,
2. Calculate the oxidation-reduction potentials of the system,
3. To determine the reaction pathway,
4. To find the EMF of the reaction, and the energy that can be produced when the electron formed by the interaction of the water molecule with the surface of ZrO_2 - nanoparticles, at $E_{V-/V} = 0.7V$; $E_{OH/OH} = 0.3V$, ΔE = EMF = 0.4V. To check the scheme 1 realization. To calculate the allocated energy W = -e ΔE [14].

1.1. The nature of the adsorption of gas molecules by the surface of nanoparticles of oxides.

Theoretically, any oxide nanoparticle is a spherical fragment 4-10 nm in diameter, isolated from a regular crystal lattice of an ionic crystal. According to the principle of electroneutrality, the surface of such a nanoparticle is initially electrically neutral, a violation of the periodicity of the crystal potential by the surface leads to the formation of additional eigen levels in the forbidden gap of the energy spectrum of the crystal. These are the so-called Shockley / Tamm localized states, which can be treated as unsaturated chemical bonds of atoms on the surface [15-17]. In the chemical sense, these are the so-called "active centers". Depending on whether the levels are donor or acceptor, the nature of the adsorbed ions will depend and, subsequently, the sign of the surface charge of the particle. In [18], it was proposed to use the value of its isoelectric point: pE as an index of the chemical activity of oxides.

Coordinative, the unsaturated surface Zr^{4+} cations are strong acid sites of adsorption (Lewis centers, pE <7), therefore, the localized states in ZrO_2 are of a donor nature. Thus, the influx of neutral molecules into the reaction zone (the region near the surface of the nanoparticles) is provided due to the individual chemical properties of the nanoparticle material and its defectiveness.

The density functional theory (DFT) calculations aimed on determining the nature of the active centers responsible for the formation of a free charge carrier in the system under study. In DFT modeling we simulate water adsorption mechanism on the surface of nanosized yttria-stabilized zirconia (YSZ) (Figure 3). For the DFT calculations we have been aimed to perform a comparative analysis of water molecule interaction with two different surfaces:

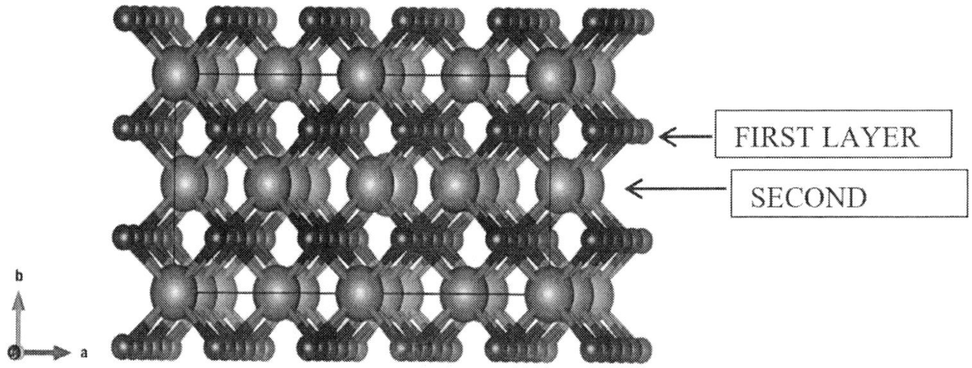

Figure 3. (Color online) A model structure of the ZrO_2 system doped with Y_2O_3. The structure model (ZrO_2+3% mol Y_2O_3) includes 15 atoms of Zr, 32 atoms O and one atom of Y. The zirconium atoms are shown in green, oxygen atoms - in red, yttrium atom - in cyan. All calculation of charge density were performed for top layers.

(1) ZrO_2 and (2) ZrO_2 + 3% mol Y_2O_3

Below the electronic structure calculations present the electronic densities, energy bands, densities of states (DOS), X-ray spectra separately for ZrO_2 and ZrO_2 + 3% mol Y_2O_3 models. Next, the DFT calculated results are compared respectively for the systems:

(3) H_2O / ZrO_2 and (4) H_2O / ZrO_2 + 3% mol Y_2O_3.

So far, the developing of computational DFT technique for the above listed molecular systems has aimed on the accurate description of their electronic structures. As one expect, the replacing of zirconium with yttrium causes a large shift in the energy gap, making the compound ZrO_2 + 3% mol Y_2O_3 a more active target for the water absorption.

1.2. The DFT technique

The density functional theory (DFT) in the Kohn-Sham formulation [19] and its practical utilization by different approaches [20] is the most widely used approach for today electronic states calculations of functional materials. In the DFT method the exchange-correlation potential is approximated by a functional of the electronic density; the most common approximations are the local density approximation (LDA) [20], the generalized gradient approximation (GGA) [21], and the hybrid approximation [22]. One have to mention some peculiarities here, that while LDA and GGA provide a successful description of ground-state properties in crystals, this success does not extend to a description of excited states. In many systems (semiconductors, crystal lattices, etc) the LDA and GGA strongly underestimate the value of the energy gap. Improved values for the band gaps are usually obtained by using the GW method [23]. However, the high computational cost of this method limits its applicability to crystals with a small number of atoms in the unit cell. An exchange potential was recently proposed by Becke and Johnson (BJ), designed to yield the exact exchange potential in atoms [24]. Unfortunately, the use of this potential led to a slight improvement in the energy gap values for many semiconductors [25]. A simple modification of the BJ potential was proposed by Tran and Blaha (TB-mBJ method). Studies have shown that the TB-mBJ potential is generally as accurate in predicting the energy gaps of many semiconductors as the much more expensive GW method [26].

In the present work, the total energy and band structure calculations are carried out using the all-electron, full potential, linear augmented plane wave (FP-LAPW) method as implemented in the WIEN2k code [27]. We have utilized the WIEN2k code calculations for the 3 mol% Y_2O_3 doped ZrO_2 interacting with H_2O. Thereby, in the DFT/WIEN2k each atom is surrounded by a muffin-tin sphere, and the total space is divided into two regions. One region consists of the interior of these non-overlapping spheres, while the rest of the space constitutes the interstitial region. The radii of the muffin-tin spheres are $2.1a_0$ for Zr, $1.9a_0$ for O, $2.1a_0$ for Y and $0.6a_0$ for H, where a_0 is the Bohr radius. In GGA calculations, the exchange correlation potential is that proposed in reference [21]. The valence electrons' wave functions inside the muffin-tin spheres are expanded in terms of spherical harmonics up to $l_{max}=10$. In the interstitial regions, they are expanded in terms of plane waves, with a wave vector cutoff of K_{max}. Because of the small muffin-tin radius of hydrogen atoms, we set $R_HK_{max}=1.12$ in ZrO_2 and ZrO_2+3% mol Y_2O_3, where $R_H=0.32a_0$ is the muffin-tin radius of the H atom. In the remaining four compounds, we set $R_{mt}K_{max}=2.1$, where R_{mt} is the smallest muffin-tin radius. The charge density is Fourier expanded up to a maximum wave vector of G_{max}, where $G_{max}=12a_0^{-1}$ for ZrO_2, $G=12a_0^{-1}$ for ZrO_2+3% mol Y_2O_3, and $G=20a_0^{-1}$ for the remaining compounds. The convergence of the self consistent calculations is achieved with a total energy tolerance of 0.001mRy and a charge convergence of 0.0004e.

2. The results summary of native experiments

2.1. The mechanism of formation of free charge carriers during the adsorption of a neutral water molecule. The potential energy of the chemical interaction of atmospheric molecules with the surface of a solid base.

Proceeding from the principle of electroneutrality [28] we will consider the number of negative and positive ions in the atmosphere equal, and the atmosphere as a whole - electrically neutral. Let us consider the mechanism of inflow to the surface of a nanoparticle based on zirconia dioxide of neutral water molecules and the selection of the energy of chemical interaction.

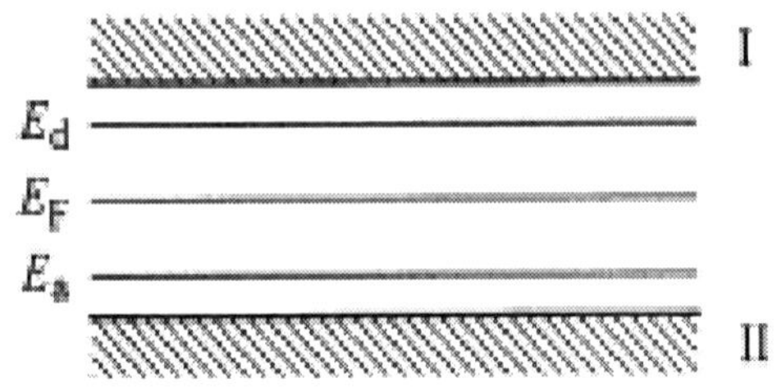

Figure 4. Qualitative band model of β-ZrO_2 with yttrium (a), with oxygen vacancies (b), with yttrium and oxygen vacancies (c).
I is the conduction band, and II is the valence band.

2.2. The nature of the adsorption of gas molecules by the surface of nanoparticles of oxides.

Theoretically, any oxide nanoparticle can be represented in the form of a spherical fragment with a diameter of 4-10 nm, isolated from a regular crystal lattice of an ionic crystal. According to the principle of electroneutrality, the surface of such a nanoparticle is initially electrically neural.

2.3. The mechanism of charge compensation of an impurity in the system ZrO_2 - Y_2O_3.
In a solid substitution solution based on zirconium dioxide, the Y^{3+} impurity atoms have one electron less than the Zr^{4+} atoms and are acceptors, that is, they create in the forbidden band levels near the top of the valence band. The oxygen vacancies V_O are formed to compensate the excess volume and charge of the impurity are electron donors, ie, donor levels (holes, near the bottom of the conduction band Figure 4) are created near the bottom of the conduction band. The acceptor and donor character of β-ZrO_2 for Y^{3+} and oxygen vacancies, respectively, is confirmed by the results of [29]. As a result of thermal fluctuations, the holes give the electron to the conduction band (ionization of the hole), which by an Coulomb interaction finds an impurity atom and hits its orbital [30]. An impurity-vacancy dipole (IVD) of the Me-V_O type is formed, and the corresponding ionic bond according to [31] is the stabilizing element of the lattice (T-phase). In this case, the impurity atom acquires a negative charge $Y^{3+(-1)}$, and the vacancy is a positive $V^{(+)}$. That is, donors and acceptors co-exist in the form of $V^{(+)}$ - $Y^{3+(-)}$IVD type[2]

2.4. The formation of an adsorption layer on the surface of particles. An own localized electronic states.
Violation of the periodicity of the crystalline potential by the surface (breaking of chemical bonds) leads to the formation of eigen levels in the band gap of the energy spectrum of the crystal-Shockley/Tamm localized states, which can be treated as unsaturated chemical bonds of atoms on the surface [32]. Their concentration in the ideal case should be equal in order of magnitude to the concentration of surface atoms. Such a configuration of the surface is not energetically favorable, so the population of these states (saturation of free valence bonds) occurs mainly by adsorption of molecules and ions from the atmosphere by the surface of nanoparticles. Thus, the influx of neutral molecules into the reaction zone (the region near the surface of the nanoparticles) is provided by the individual chemical properties of the nanoparticle material and its defectiveness.

2.5. Adsorption of molecules from the gas phase in ZrO_2 - 3 mol% Y_2O_3.
If the level is acceptor, on it with a probability determined by the Fermi-Dirac function

$$f = \left[1 + \exp\left(E_f - E_c\right)/kT\right]^{-1} \quad (1)$$

The electron (belonging to the nanoparticle material) is localized, additional surface states are filled at the bottom of the conduction band, and the bond formation between the surface and adsorbates is formed by the exchange interaction. As a result, a potential-forming layer of adsorption origin is formed (Helmholtz layer [33, 34], 2, Figure 5), and the neutral surface acquires a charge. In the case of ZrO_2 - 3 mol% Y_2O_3, this charge is negative. To compensate the charge of the potential-forming layer, an external diffuse layer is formed (3, Figure 5).

2.6. The structure of the reaction zone.
A particle of a dispersed phase, together with a DEL, is called a micelle (Figure 5). The aggregate of the substance {[ZrO_2] m} with potential-forming ions, (predominantly OH- for ZrO_2 + 3 mol% Y_2O_3 [35, 36]) form the core of the micelle {m [ZrO_2] nOH^-}. Potential-forming ions are connected with the surface of a relatively strong (E> 0.3 eV [37]) chemical bond. The diffuse layer is connected with the core of the micelle physically by the forces of electrostatic interaction and consists mainly of H^+

2 The nanoparticle as a physical object can be considered theoretically as a nano-sized fragment of an extended electrically neutral crystal lattice of a solid solution of the appropriate composition. Consequently, initially the nanoparticle is electrically neutral.

protons and polarized water molecules. It is the diffuse layer that is the receptor of molecules from the outer atmosphere and their supplier into the reaction zone. The system of adsorbates on the surface of ZrO_2 is in dynamic equilibrium [38]. Consequently, when the external conditions change, it undergoes a change and a charge state of the near-surface layer.

The real mechanism of interaction of the water molecule with the surface of the micelle is multistage and is theoretically considered in detail in [39-40]. An interesting question is how the adsorption energy is transferred from the diffuse layer to the heterophase boundary, that is, to the surface of the nanoparticles.

2.7. The physical basis of the chemoelectronic effect.

Upon reaching the surface of an electrically neutral zirconium oxide nanoparticle, the neutral water molecule under dissociation of the surface force gradient dissociates into a proton of the H^+ and OH-group. In the case of physical adsorption, an electrochemical process is realized such as [41]:

$$H_2O + e^- \rightarrow H_{ads} + OH^-$$

The total charge of the surface changes, in this case, becomes negative. This leads to bending of the levels in the energy bands, and the charge exchange of electronic states localized near the surface. In particular, the localization of a part of electrons from the crystal lattice of nanoparticles near the surface and the formation of a space-charge region of an electronic type occur there with the probability determined by the Fermi-Dirac function. This region can be used as electrical capacitance in the development of ionistors.

In the case of chemical adsorption, processes with electron transfer via the phase interface are probably realized. Taking into account the acceptor character of the impurity level in the system under study, it can be assumed that the process proceeds according to the scheme [42]:

$$H_2O + e^- \rightarrow H_2O^- \rightarrow H^{\bullet} + OH^-,$$

where $H^{\bullet}$ – a radical hydrogen form.

Schematic representation of the proposed process is shown in Figure 6. The formation of a radical form of hydrogen leads to the launch of a chain reaction of ionization of the molecules of the dispersion medium. As a result, by the relay mechanism the charge is transported to the region with a lower charge concentration of the corresponding sign.

Unlike photocatalytic systems, the nanopowder heterophase system is thermodynamically closed, i.e., does not receive energy from outside as radiation, and as a consequence has a specific physical limit on electron production.

3. Simulation results

3.1. ZrO_2 : crystal and energy bands structures.

The cubic unit cell with space group F m -3 m and lattice constants a=b=c=4.938 Å was used in DFT calculation. The lattice parameters obtained (Table 1.) show excellent agreement with the experimental results and the differences is less than 3%. The calculated surface energies are presented in Table 2, which the minimum lattice energy in the 120.13 Å^3 is received -165.2 eV. This volume of unit cell was chosen to all calculation experiential. Calculations were performed using Abinit code and GGA (PBE) - Fritz-Haber-Institute (FHI) pseudopotential with energy convergence of 0.0001.

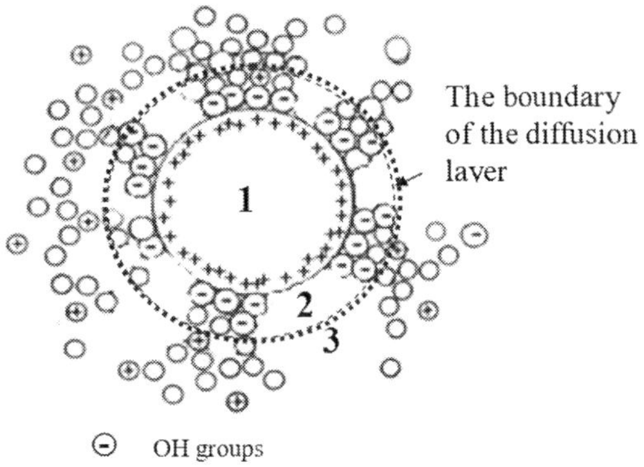

Figure 5. Schematic representation of a ZrO_2 (1) nanoparticle surrounded by an ionic atmosphere, including adsorption (2) and diffuse (3) layers.

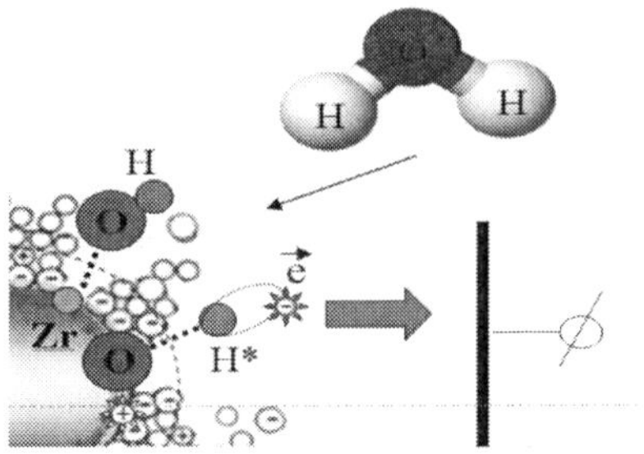

Figure 6. Schematic representation of a possible electrochemical process with an electron transfer through the phase-gap interface during adsorption of water molecules to the surface of nanoparticles based on ZrO_2.

Table 1. Calculated and observed structural parameters of cubic ZrO_2 (a[43]; b[44]; c[45]).

Parameters	(exp)	(calc)
a=b=c (Å)	4.913[a] 5.07[b]	4.938
Lattice energy per ZrO_2 (eV)	-109.76[c]	-165.2
Lattice Energy per ZrO_2+3% mol Y_2O_3 (eV)	-106.1[c]	141.2

Table 2. Lattice energy per unit cell for both ZrO_2 and ZrO_2+3% mol Y_2O_3.

Lattice Energy per unit cell ZrO_2 (eV)	-138.6	-158.4	-163.1	-159.25	-165.2	-148.13	-84.1
Lattice Energy per unit cell ZrO_2+3% mol Y_2O_3 (eV)	-124.75	-137.88	-139.1	-137.25	-141.2	-132.75	-76.75
Volume $Å^3$	114.79	116.58	118.58	119.35	120.13	121.43	122.46

In Figure 7 we present the calculated energy bands of ZrO_2. The electronic density of states in Figure 7 have shown for high symmetry directions of the Brillouin zone (BZ). The valence band maximum (VBM) and conduction band minimum (CBM) occur at the Γ-point, the BZ center. In ZrO_2, the gap occurs at point R(1/2, 1/2, 1/2). Point R of the cubic lattice BZ is zone-folded into the Γ-point of the body-centered tetragonal lattice BZ. The density of states of ZrO_2 is shown in Figures 8 (top and bottom), where we see that the low-lying conduction bands are derived from O *p* states. On the other hand, the bands in the range -2 eV to 0 eV are dominated by oxygen-derived states. The valence band just below the Fermi energy is derived from zirconium *s* and *p* states. These observations become clear upon considering the atomic orbital character of the bands, which is presented in Figure 8 (bottom). The contribution of the chosen atomic orbital to the eigenstates at each k-point, where the CBM is derived mostly from Zr *p* states. The VBM also dominated by Zr, though a mixture of Zr *s* and O *p* states is clear. The anti-bonding state formed from these *s* and *p* states is pushed up in energy close to the Fermi level. The large contribution of Zr s ($l = 0$) states to the VBM and Zr *p* ($l = 1$) states to the CBM suggests that there are some transitions between the VBM and CBM ($\Delta l = 1$), to be useful of this material in charge conduction process.

3.2. ZrO_2+ 3% mol Y_2O_3 : crystal and energy bands structures.

A model structure of the zirconium dioxide doped with yttrium is shown in Figure 9. The structure model (ZrO_2+3% mol Y_2O_3) includes 15 atoms of Zr, 32 atoms O and one atom of Y. The zirconium atoms are shown in green, oxygen atoms - in red, yttrium atom - in cyan. All calculation of charge density were performed for top layers as shown in Figure 3 above.

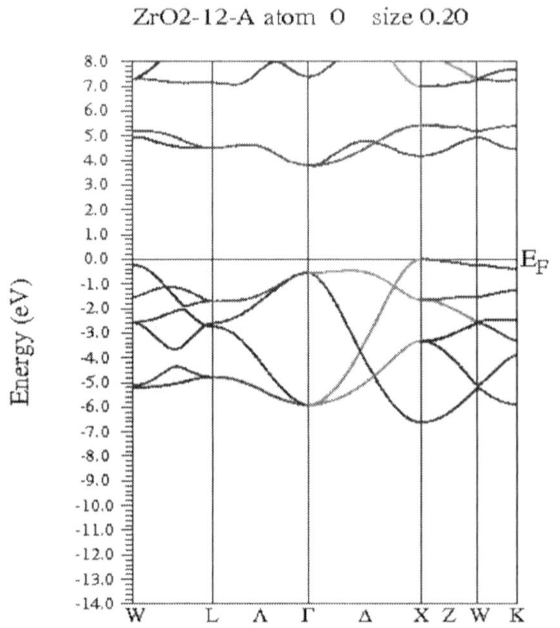

Figure 7. The calculated energy bands of ZrO_2.

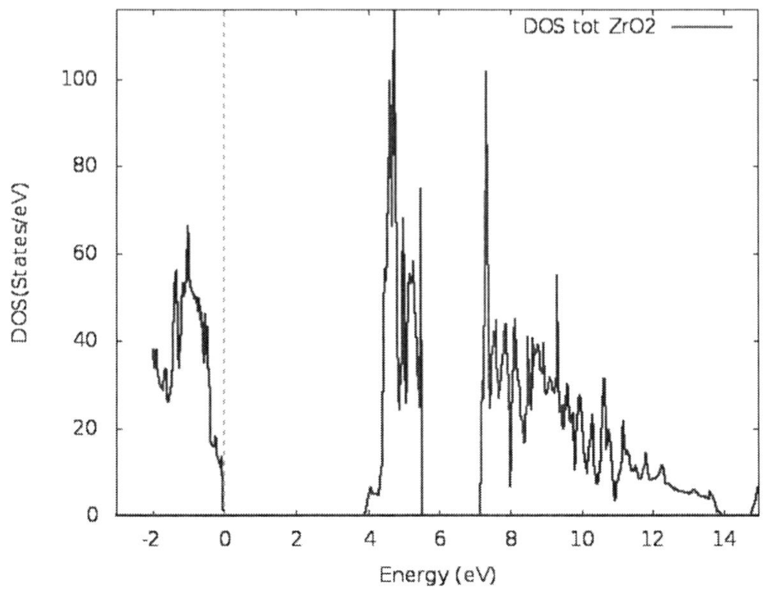

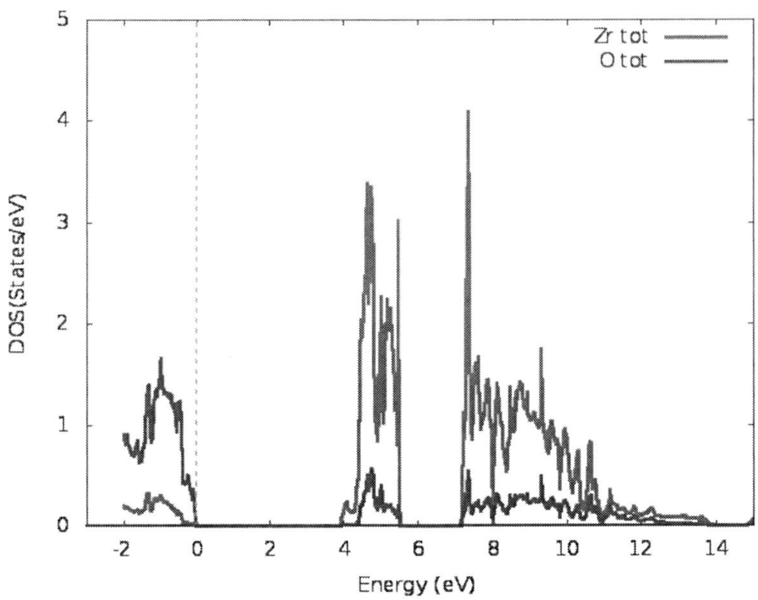

Figure 8. The calculated densities of states of ZrO_2 (top and bottom).

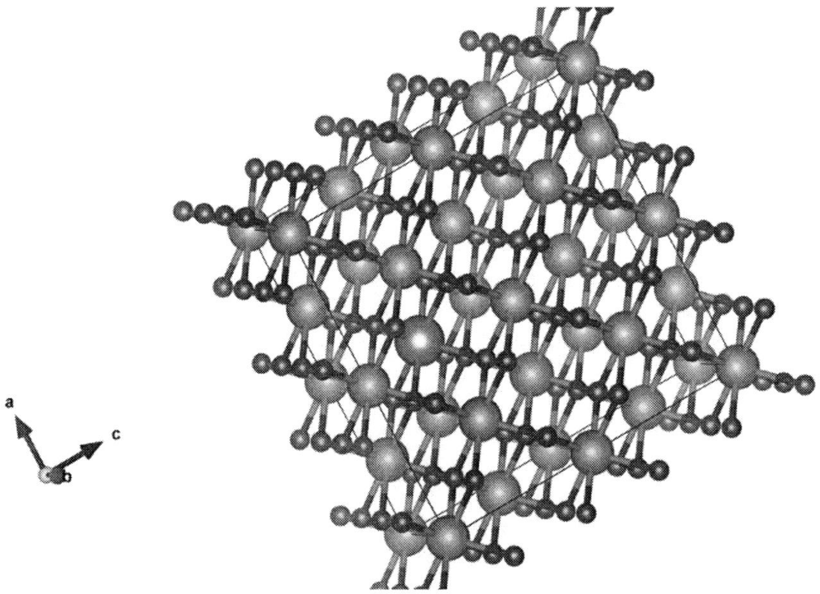

Figure 9. (Color online) A model structure of the ZrO_2 system doped with Y_2O_3. The zirconium atoms are shown in green, oxygen atoms - in red, yttrium atom - in cyan.

In Figure 10 the calculated energy bands are presented for the system ZrO_2 + 3% mol Y_2O_3. The electronic density of states in Figure 11 demonstrate the relatively higher densities, in comparison with ZrO_2 (Figure 7), and narrowing of the gap between the VBM and CBM zones. The Fermi level slightly shifted up to, so the gap between the VBM and CBM has narrowed on 1 eV.

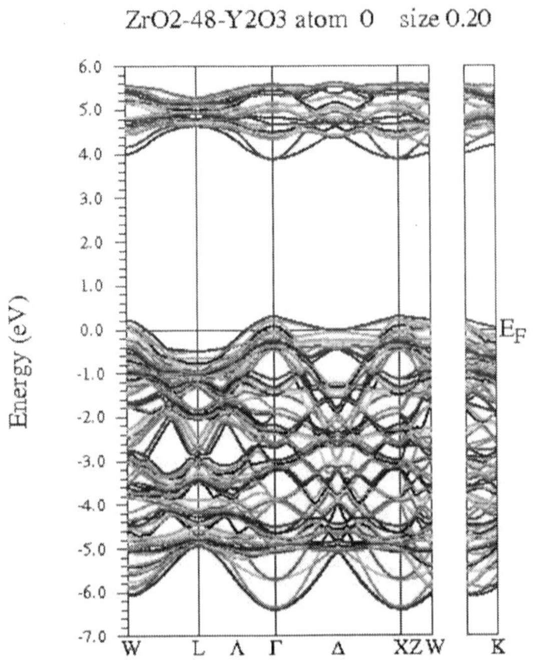

Figure 10. The calculated energy bands of ZrO_2+ 3% mol Y_2O_3.

Figures 11 (top and bottom) show the density of states for ZrO_2 + 3% mol Y_2O_3. Now doping of yttrium to the zirconia structure causes the dominating of oxygen in the interval from -5 eV to 0 eV. At the same time, the yttrium with zirconium dominate together in the conduction zone. These observations clearly demonstrate the atomic orbital modifications of the bands, which are influenced by the yttrium doping to the zirconium dioxide.

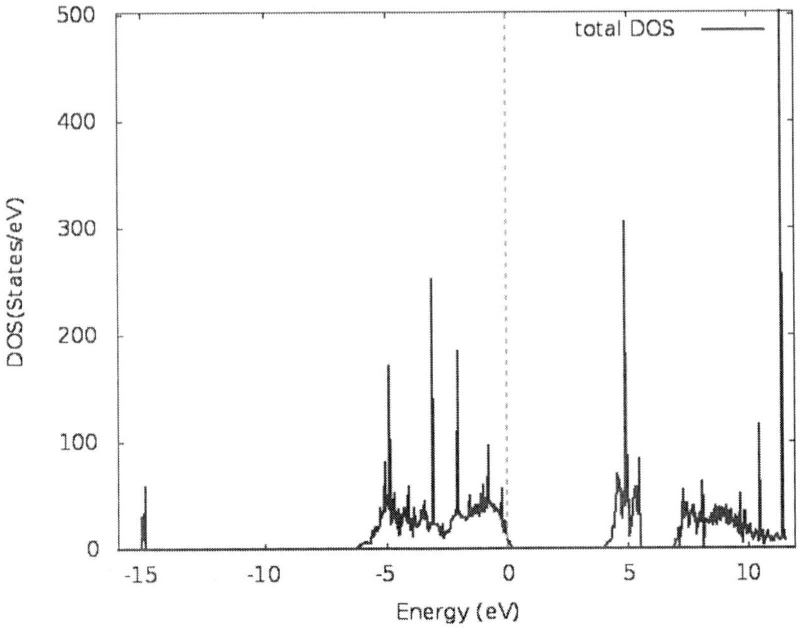

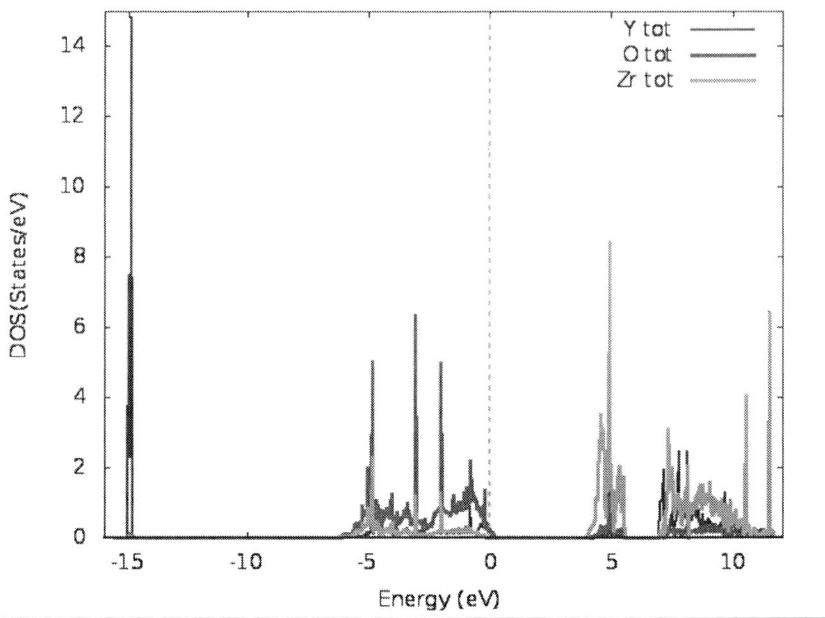

Figure 11. The calculated densities of states of ZrO_2 + 3% mol Y_2O_3 (top and bottom).

The calculated above electronic densities of ZrO_2 + 3% mol Y_2O_3 should be compared with the ZrO_2 model. We can observe the DOS spectra peculiarities for ZrO_2 + 3% mol Y_2O_3 that possess visibly higher amplitudes in comparison with ZrO_2 ones. This observation could be a precursor for a charge re-distribution process, anticipating a charge conduction of agents as like as water molecule on the ZrO_2 + 3% mol Y_2O_3 surface.

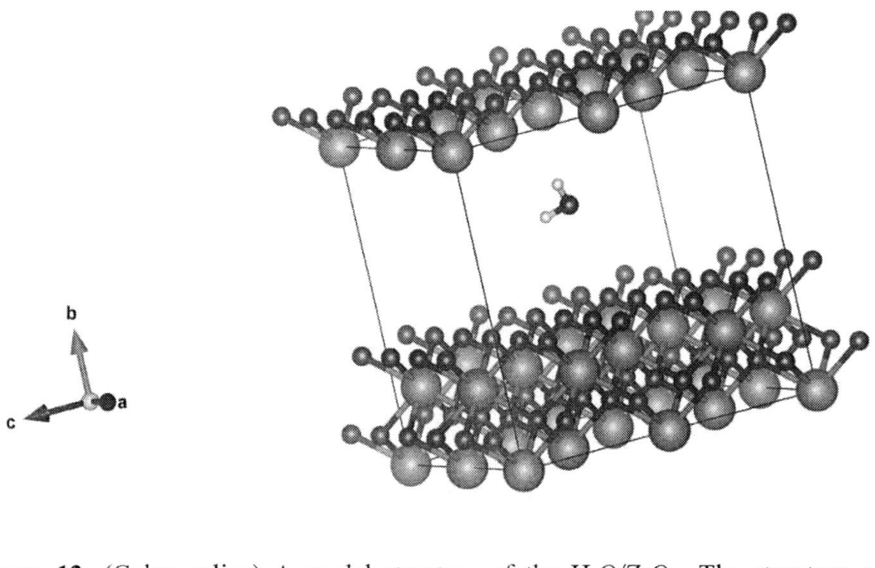

Figure 12. (Color online) A model structure of the H_2O/ZrO_2. The structure model (ZrO_2+H_2O) includes 16 atoms of Zr and 32 atoms O of the periodic surface, and one water molecule. The zirconium atoms are shown in green, oxygen atoms - in red, hydrogen atom - in white. The distance between atom of oxygen of H_2O and first layer is 6.334199 Bohr.

3.3. H_2O/ZrO_2 – system.

So far, the use of DFT for the above listed molecular systems has aimed on the accurate description of their comparative electronic structures. As one expect, the replacing of zirconium with yttrium causes a large shift in the energy gap, making the compound ZrO_2 + 3% mol Y_2O_3 a more active target for the water absorption. Next, the DFT calculated results are compared respectively for the systems: (3) H_2O / ZrO_2 (Figure 12) and (4) H_2O / ZrO_2 + 3% mol Y_2O_3 (Figure 14).

In Figures 12-13 and 14-15 the configuration snapshots and calculated DOS are shown for H_2O/ZrO_2 and H_2O/ZrO_2+ 3% mol Y_2O_3, respectively. The important observation is that the inducing 3% mol Y_2O_3 doping to system H_2O/ZrO_2 brought to the shift of DOS amplitude from 4 eV to 0. This is obviously indicate on a charge redistribution of the water interaction with zirconia surface as a result of ittrium ion doping.

Concluding the DFT results with the experimental data, it is worth noting that the idea is that, when molecule of water approaching on the surface of ZrO_2 (ZrO_2+3% mol Y_2O_3) it's kinetic energy leads to a change the DOS and the electron density of the ZrO_2 (ZrO_2+3% mol Y_2O_3). On the basis of the first Hohenberg and Kohn theorem the total energy calculated with this density of a electron system $\rho(r)$ will always be larger or equal then the ground state energy E_0

$$E_0 \leq \mathrm{E}[\rho(r)] \tag{2}$$

At the kinetic energy transition of the water molecule T to above system (ZrO_2) according the first Hohenberg theorem the electron density of the system should be changed:

$$E_0 + T \leq \mathrm{E}[\rho(r) + \Delta\rho(r)] \tag{3}$$

Depending on the distance between water molecule and first layer of ZrO_2 (ZrO_2+3% mol Y_2O_3) the kinetic energy transition will be different. Hence, in this simulation the distances between the water molecules and the oxide surface was chosen for the 4 values.

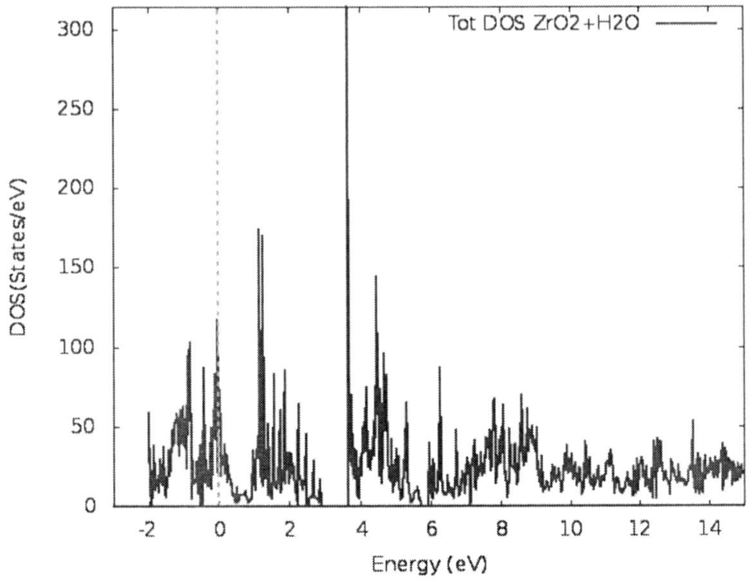

Figure 13. The calculated densities of states of H_2O/ZrO_2 (top, middle and bottom). The distance between atom of oxygen of H_2O and first layer is 2.6205095 Bohr.

3.4. H_2O/ZrO_2+ 3% mol Y_2O_3- system

Concluding the DFT results with the experimental data, it is worth noting that the idea is that, when molecule of water approaching on the surface of ZrO_2 (ZrO_2+3% mol Y_2O_3) leads to a

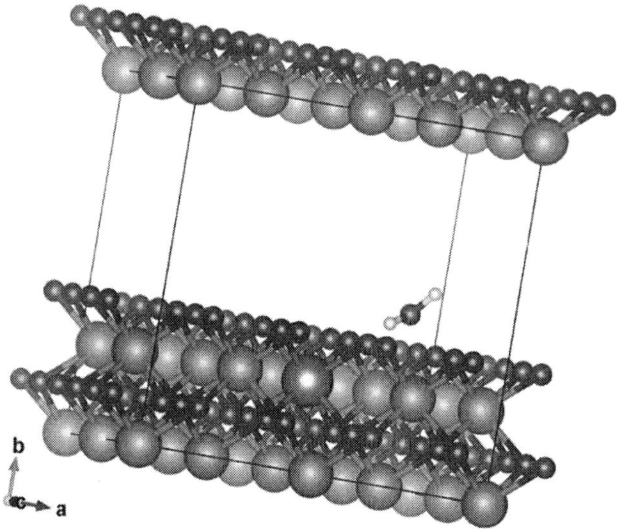

Figure 14. (Color online) A model structure of the H_2O/ZrO_2+ 3% mol Y_2O_3. The structure model (H_2O/ZrO_2+ 3% mol Y_2O_3) includes 15 atoms of Zr, 1 atom Y and 32 atoms O of the periodic surface, and one water molecule. The zirconium atoms are shown in green, oxygen atoms - in red, hydrogen atom - in white and yttrium atom - in cyan. The distance between atom of oxygen of H_2O and first layer is 1.32077 Bohr.

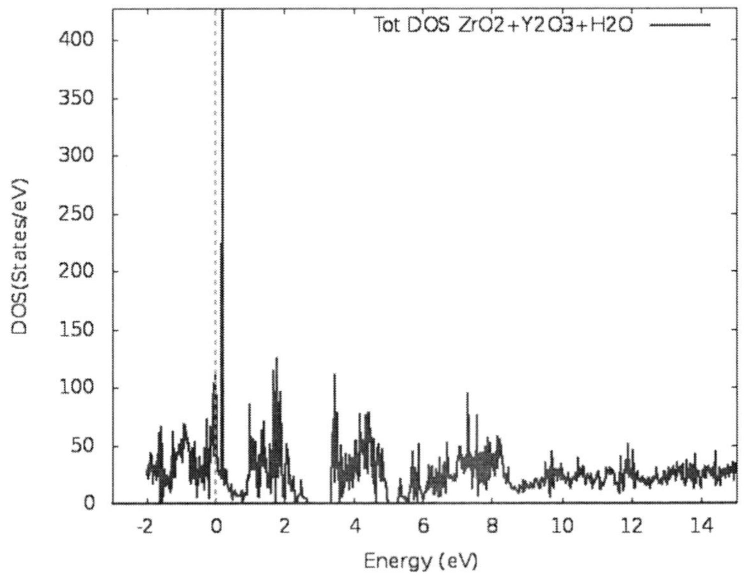

Figure 15. The calculated densities of states of H_2O/ZrO_2+ 3% mol Y_2O_3 (top and bottom). The distance between atom of oxygen of H_2O and first layer is 2.6205095 Bohr.

change the DOS and the electron density of the ZrO_2 (ZrO_2+3% mol Y_2O_3). On the basis of the first Hohenberg and Kohn theorem [46] the total energy calculated with this density of a electron system ρ(r) will always be larger or equal then the ground state energy $E_0, E_0 \le \mathrm{E}[\rho(r)]$. At the kinetic energy transition of the water molecule T to above system (ZrO_2) according the first Hohenberg theorem the electron density of the system should be changed: $E_0 + T \le \mathrm{E}[\rho(r) + \Delta\rho(r)]$. Depending on the distance between water molecule and first layer of ZrO_2 (ZrO_2+3% mol Y_2O_3) the kinetic energy transition will be different. Hence, in this simulation the distances between the water molecules and the oxide surface was chosen for the 4 values.

4. Conclusions

Density functional theory calculations have been performed on ZrO_2 nanoparticles doped with Y_2O_3 and interacting with the water molecule to evaluate the electronic density, total energy and band structures. It is shown that the surface of ZrO_2 nanoparticles can be considered as a reaction zone for electrochemical processes.

The important observation is that the inducing 3% mol Y_2O_3 doping to system H_2O/ZrO_2 brought to the shift of DOS amplitude from 4 eV to 0. It's show's that water molecules adsorbed from the atmosphere on the surface of ZrO_2-based nanoparticles leads to realization of the process of electron localization from the crystal lattice of nanoparticles and its transport outside the particles.

Acknowledgments
RFBR grant No.17-52-45062 ИНД_a is acknowledged.
A.S. Doroshkevich, A.I. Lyubchyk acknowledges funding by the EU H2020-MSCA-RISE-2015 through the HUNTER project (grant nº 691010).
JINR-Romania Cooperation Programme Project of 2017 Order No. 219 / 55 is acknowledged.

References

[1] Doroshkevich A S, Lyubchyk A I, Islamov A Kh at all 2017 Computer Design for New Drugs and Materials: Molecular Dynamics of Nanoscale Phenomena 10 139
[2] Voznaya and Water N. F 1979 Chemistry and Microbiology (Moscow: Higher School) p 340
[3] Styrov, V V, Simchenko, S V 2013 Reports of the Nat. Acad. of Sci. of Ukraine 80 5
[4] Styrov V V, Tyurin Yu I, 2003 Non-equilibrium chemo-effects on the surface of solids (Moscow: Energoatomisdat).
[5] Georgen B, Nienhaus H, Weinberg W H, McFarland E. 2001 Science 294 2521
[6] Fujishima A and Honda K 1972 Nature 238 37
[7] Kabanskii A Y, Styrov V V 2004 Advanced Materials for Energy Conversion II USA: Publ. TMS, 43.
[8] Kabanskii A E, Yu.I. Tyurin 1979 JETP 5 14 (in Russian)
[9] Styrov V V and Simchenko S B 2012 JETP 5-6 96 (in Russian)
[10] Nenad Miljkovic, Daniel J Preston, Ryan Enright, Evelyn N Wang 2014 Applied Physics Letters 105 013111
[11] Mingming Ma, Liang Guo, Daniel G Anderson and Robert Langer 2013 Science 339 6116
[12] Doroshkevich A S, Lyubchyk A I, Shilo A V, Zelenyak T Yu, at all 2017 Journal of Surface Investigation: X-ray, Synchrotron and Neutron Techniques 3 11
[13] Doroshkevich A S, Lyubchyk A I, Shilo A V, Zelenyak T Yu, at all 2017 IOP Conf. Series: Journal of Physics: Conf. Series 848 012021
[14] http://5terka.com/node/495)
[15] Koutecky J 1957 Phys. Rev. 1 10
[16] Tamm I Y 1933 JETP 3 11 (in Russian)

[17] Shokley W 1939 Phys. Rev. 1 59
[18] Shukla A K Manoharan R, Goodenough J B 1988 Solid State Ionics 26 5
[19] Kohn W, Sham L J 1965 Phys. Rev. 140 1133
[20] Koller D Tran; F, Blaha P 2011 Phys. Rev. B 83 195134
[21] Perdew J P Burke K 1996 Phys. Rev. Lett. 77 386
[22] Becke A D 1993 J. Chem Phys. 98 1372
[23] Bechstedt F Fuchs,F, Kresse G 2009 Phys. Status Solid B 246 1877
[24] Becke A D Johnson E R 2006 J. Chem. Phys.124 221101
[25] Tran F Blaha P, Schwarz K 2007 J. Phys.: Condens. Matter 19 19620
[26] Tran F, Blaha P 2009 Phys. Rev. Lett. 102 22640
[27] Blaha P , Schwarz K, Madsen G K H, Kvasnicka D, Luitz J 2001 WIEN2K: An Augmented Plane Wave and Local Orbitals Program for Calculating Crystal Properties (Vienna: University of Technology Institute of Materials Chemistry)
[28] Iossel YU YA, Klenov G E 1984 Matematicheskiye metody rascheta elektrokhimicheskoy korrozii i zashchity materialov Sprav Izd. (Moskva: Metallurgiya) 272 s. (in Russian)
[29] Naymov I I at all 1992 Inorganic Materials 4 28 (in Russian)
[30] Uert Ch, Tomson R 1969 Fizika tvordogo tela (Moskva: Mir) 280 s. (in Russian)
[31] Alekseenko V I Volkova G K 2000 Technical Physics Letters 70 9 (in Russian)
[32] Koutecky J 1975 Phys. Rev. 10 1
[33] Voyutskiy S S 1975 Kurs kolloidnoy khimii (Moskva: Khimiya) 512s. (in Russian)
[34] Fridsberg D A 1984 Kurs kolloidnoy khimii: [ucheb. dlya vuzov] / D.A. Fridsberg. – [2-ye izd., pererab. i dop.]. (Leningrad: Khimiya) 368 s. (in Russian)
[35] Blyumental U B 1963 Khimiya tsirkoniya. Pod redaktsiyey Komissarovoy L N i Spitsyna V I (Moskva: Izd-vo inostrannoy literatury) 345 s. (in Russian
[36] Lisichkin G V 1996 Soros Educational Journal 4 52 (in Russian)
[37] Akopyan M E 1998 Soros Educational Journal 2 115 (in Russian)
[38] Yasuharu O 2008 Applied Surface Science 255 3434
[39] Kvlividze V I 1970 Izucheniye adsorbirovannoy vody metodom yadernogo magnitnogo rezonansa // Svyazannaya voda v dispersnykh sistemakh. (Moskva: MGU) 1 41 (in Russian)
[40] Ignatyeva L A, Kvlividze V I, Kiselev V F 1970 O mekhanizme elementarnogo akta vzaimodeystviya vody s poverkhnost'yu okislov // Svyazannaya voda v dispersnykh sistemakh. (Moskva: MGU) 1 56 (in Russian)
[41] Vanmaekelbergh D, Houtepen A J, Kelly J J 2007 Electrochimica Acta 53 1140
[42] Kudryashov YU B 2004 Radiatsionnaya biofizika (Moskva: Fizmatlit) s. 208. (in Russian)
[43] Wu H, Duan Y, Liu K, Ly D, Qin L, Shi L,Tang G 2015 Journal of Alloys and Compounds 645 352
[44] Catlow C R A 1990 J. Chem. Soc., Faraday Trans. 86 1167
[45] Xin X 2010 Computational Modelling Study of Yttria-stabilized Zirconia, Thesis submitted for the degree of Doctor of Philosophy (London: Department of Chemistry University College.) p 217.
[46] Hohenberg P and Kohn W 1964 Phys. Rev. B 136 864.

The influence of high pressure to crystalline and magnetic structure of Ba_2FeMoO_6

V A Turchenko[1,2], N A Kalanda[3], L V Kovalev[3], M V Yarmolich[3], A V Petrov[3], Ye V Lukin[1], A S Doroshkevich[1,2], M Balasoiu[1,4], N Lupu[5] and B N Savenko[1]

[1] Frank Laboratory of Neutron Physics, Joint Institute for Nuclear Research, 6 Joliot-Curie str., 141980 Dubna, Russia
[2] Donetsk Institute of Physics and Technology named after O.O. Galkin of the NASU, 46 Nauki Ave, 03680, Kiev, Ukraine
[3] SSPA Scientific and practical materials research centre of NAS of Belarus, 19 P. Brovki str., 220072 Minsk, Belorussia
[4] Horia Hulubei National Institute for Physics and Nuclear Engineering, Bucharest, Romania
[5] National Institute of Research and Development for Technical Physics, Iasi, Romania

E-mail: turchenko@jinr.ru

Abstract. The behavior of the crystalline and magnetic structure of Ba_2FeMoO_6 compound in a wide pressure range from 0 to 4.7 GPa was studied. The crystal structure of ceramic sample was described in the framework of SG I4/mmm (No 139) and contains less 10% of anti-site defects. The change of tetragonal structure (I4/mmm) was not observed in all measured pressure range. It was shown multidirectional influence of ambient pressure onto the average interionic distances of metal-ligand in oxygen octahedrons of FeO_6 and MoO_6. For tetragonal structure of Ba_2FeMoO_6 were determined coefficients of the linear and all-round compressibility. The influence of ambient pressure on the value of magnetic moment of iron sublattice was shown.

1. Introduction

Oxides of double perovskites $A_2MM'O_6$ (where A = Ba, Sr, Ca ...; M = Fe, Cr ...; M' = Mo, W, Re ...) and their solid solutions attract the attention of researchers due to variety of physical properties associated with chemical composition and ordering of cations in crystalline structure. In general, cations M and M' can be distributed randomly in their sublattices, however, if there are a large difference in charges or ionic radii between them, they can arranged ordered. In this case, the crystal structure of these materials contains arranged in staggered order oxygen octahedrons MO_6 and $M'O_6$, where M and M' are cations of different chemical elements [1]. However, an appearance of anti-site defects (ASD) in crystal structures of these materials is able significantly influence onto their functional properties. This sort of structural disorder occurs when M ions are located at M' sites and vice versa. Several earlier theoretical and experimental studies have shown clear connection between the magnetic properties of SFMO and the anti-site disorder. The saturation magnetization [2, 3, 4, 5] and the Curie temperature [3, 6] has been seen to decrease with increasing amount of ASD. In addition to the T_C and M_s ASD reduces the band gap in the majority spin channel and this effect is large enough that the half metallic feature of SFMO is lost [7, 8, 9]. Double perovskites have higher values of the Curie temperature [10, 11, 12] and their tunnel type magnetoresistive properties appear in a weak range of magnetic fields near the room temperature [12, 13], as opposed to ordinary manganite-lanthanum perovskites. Another feature of this materials is there ferromagnetic half metallic properties, which have a potential for application in electronic devices of spintronics [12, 14]. The spin polarization of carriers controlled by the magnetic field will determine the double perovskite as a conductor for electrons with spin oriented upwards (or down), or as an insulator for the other charge carriers with an oppositely directed spin orientation. It is possible to influence onto the spin polarization of carriers of double perovskites by controlling the concentration of ani-sites defects or creating "chemical presse", for example the incommensurate substitution of ions in the $A-$ sublattice influences to bond-lengths and angles among ions of metal and oxygen and, consequently, onto the orbital overlap of ions located at M and M' sites. Another way to control the spin polarization of charge carriers of these materials is the external action, for example an ambient temperature and pressure. The main aim of this work is ascertain regularities of influence of high pressure (up to 4.7 GPa) on the features of the crystalline and magnetic structure of the double perovskite Ba_2FeMoO_6. A main feature of this work is application of the neutron diffraction method that allowed one to collect information about both the crystalline and magnetic structure of the sample during one experiment and over a wide pressure range (from 0 to 4.7 GPa).

2. Methods and methodology

The ceramic sample of Ba_2FeMoO_6 has been fabricated by solid state method from a mixture of Fe_2O_3 and MoO_3 oxides of analytical grade purity and $BaCO_3$ carbonate taken in appropriate ratios. The initial mixture was synthesized at 900^oC (4 h) in air and then was annealed at 1200^oC (10 h) under H_2/Ar stream. After annealing, the sample was slowly cooled ($\sim$ 100 ^oC/h). The phase composition was determined by X-ray diffraction method using an Empyrean diffractometer (firm PANalytical) in Cu-K_α radiation at room temperature. Structural studies at high pressures (up to 4.7 GPa) have been performed using a high-pressure chamber with sapphire anvils on a DN-6 neutron diffractometer [15] (Dubna, Russia). An 1 mm^3 sample was placed in an anvil hole drilled in the center, that allowed to create a quasi-hydrostatic pressure distribution on the sample. The unevenness of pressure distribution across the surface of the test sample in the anvils with the holes usually does not exceed 15%. The magnitude of the applied pressure was measured from the shift of the ruby fluorescence line [16] (doublet 6942 and 6927 Å) to within 0.05 GPa. The neutron diffraction spectra were measured at scattering angles $2\Theta = 90^0$ and 45^0. For these scattering angles, the resolution of the diffractometer at wave length $\lambda = 2$ Å was $\Delta d/d_0 = 0.02$ and 0.025, respectively. A typical measurement time of one spectrum was 12 hours. Analysis of neutron diffraction patterns with the Rietveld method has been carried out by the software package FullProf [17] using the built-in tables for coherent scattering lengths and magnetic form factors. Measurements of the magnetization of the Ba_2FeMoO_6 sample in the temperature range 77 – 900 K and in a magnetic field of 0.86 T were carried out using an universal automated appliance.

3. Results and discussions

3.1. Magnetic properties

The behavior of the specific magnetization shown in Fig.1 characterizes the response of magnetic properties of Ba_2FeMoO_6 onto a change of ambient temperature. The perovskite sample is in a saturated state at low temperatures. A gradual increase the ambient temperature leads to a violation of the long-range magnetic ordering in the sample because of the increasing of thermal motion of ions and the magnetization is decreased. When the temperature reaches $T_C \sim 317$ K the energy of thermal motion becomes comparable with the energy of the exchange interaction, as a consequence the magnetic order is destroyed and the sample of Ba_2FeMoO_6 passes from the magnetically ordered to the paramagnetic state. The Curie temperature was determined as the inflection point in the temperature dependence of specific magnetzation.

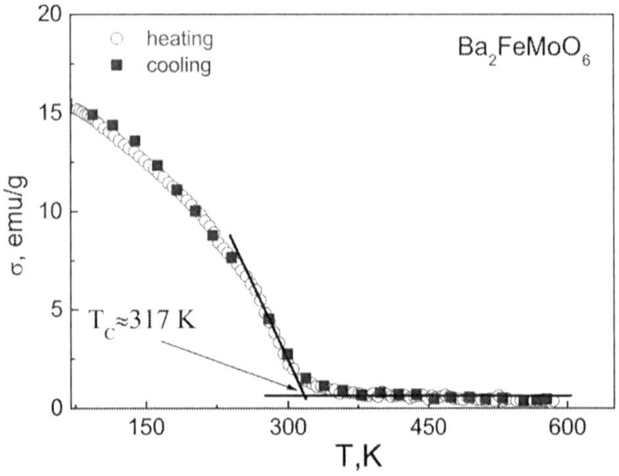

Figure 1. The temperature dependence of speific magnetization of the Ba_2FeMoO_6 compound.

3.2. Crystal structure

According to X-ray diffraction data, the sample Ba_2FeMoO_6 is homoheneous. An analysis of the crystal structure of Ba_2FeMoO_6 has been carried out by the Rietveld method. The results of the analysis are shown in Fig.2. At room temperature, Ba_2FeMoO_6 has a tetragonal structure with the sp. gr. I4/mmm (No. 139). The scheme of the unit cell of the tetragonal structure is shown in Fig.3, a.

An important characteristic that determines the functional properties of double perovskites is the concentration of ASD which can be determined from the ratio of integrated intensities of the diffraction peaks: $I_{(101)}/(I_{(200)} + I_{(112)})$ [18]. It should be noted that the diffraction peak (101) in the double perovskite structure is superstructural and and it is associated with the Fe/Mo alternative ordering. In completely disordered structure when a half of iron ions occupy positions in the molybdenum sublattice and vice versa, the diffraction peak (101) is absent. Examples of X-ray patterns corresponding to fully ordered and disordered structures were simulated in the FullProf program and they are shown in the inset of Fig.2. The degree of structural disorder was determined, similarly to the procedure described in [19]:

$$R = 0.000271 \cdot (ASD)^2 + 0.000684 \cdot (ASD) - 0.0035 \quad (1)$$

According to Eq.1, the concentration of anti-site defects (ASD) in Ba_2FeMoO_6 is less 10%.

3.3. Crystal structure under high pressure

The investigation of the atomic structure under the influence of high pressures allows ascertaining the relationship among changes of structural parameters, interatomic distances, magnetic structure and macroscopic properties that is necessary for understanding the nature and mechanisms of physical phenomena of double perovskites. Examples of neutron diffraction patterns Ba_2FeMoO_6 obtained at a temperature of 290 K in the range of applied pressures from 1 to 4.7 GPa are shown in Fig.4. The presence of additional superstructural peaks on

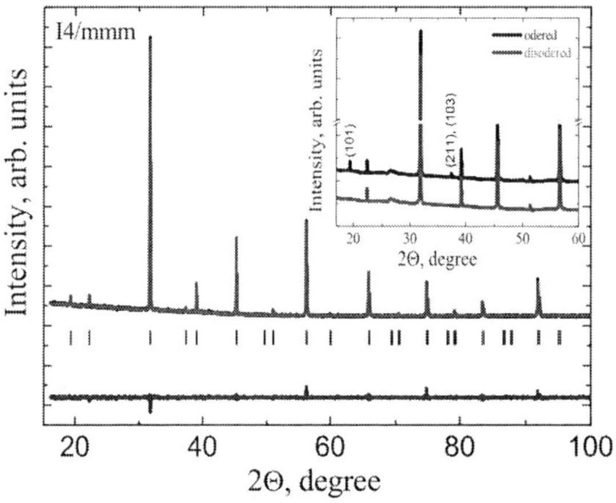

Figure 2. X-ray diffraction pattern of Ba_2FeMoO_6, measured at 290 K and fitted by the Rietveld method. The experimental and calculated values (upper curves) and a difference curve (the bottom line) normalized to a statistical error are presented. Vertical bars are calculated positions of diffraction peaks corresponding to the crystal structure of Ba_2FeMoO_6 (sp. gr. I4/mmm) at 290 K. The inset shows X-ray patterns for completely ordered (upper curve) and disordered (lower curve) crystal structure of the Ba_2FeMoO_6 simulated in FullProf software.

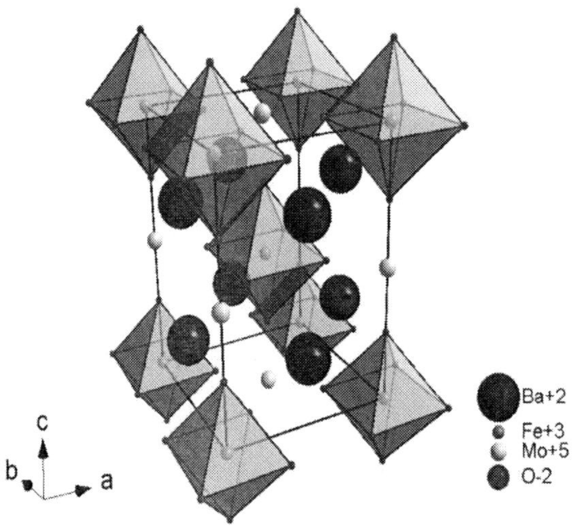

Figure 3. Scheme of the tetragonal crystal structure of Ba_2FeMoO_6 at room temperature.

neutron patterns of Ba_2FeMoO_6 in the entire range of applied external pressures was not observed that indicates the absence of a change in the type of crystal and magnetic structure.

An increase of the external pressure leads to a monotonic decrease in the parameters of the unit cell (see Fig.5).

The results of refining by the Rietveld method of neutron diffraction patterns of Ba_2FeMoO_6 at different external pressures are shown in Table 1. The dependences of volume of unit cell

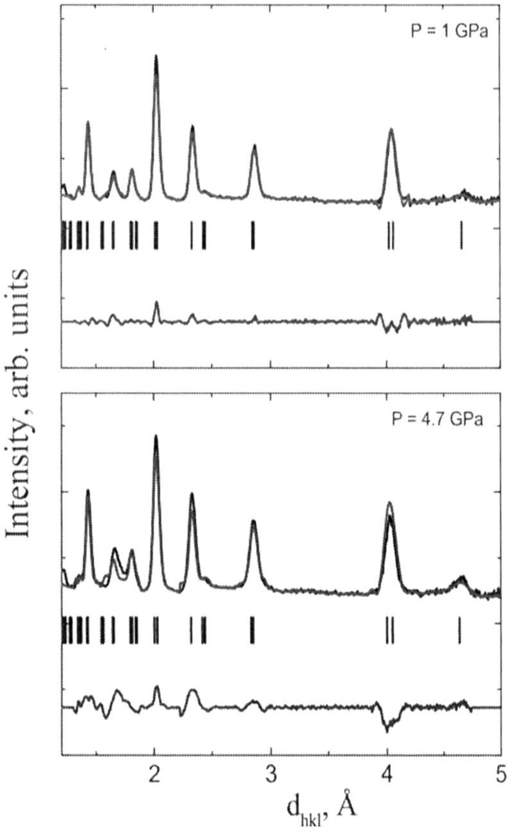

Figure 4. Neutron diffractograms of Ba_2FeMoO_6, measured at different pressures and processed by the Rietveld method. Experimental and calculated values (upper curves) and a difference curve (the bottom line) normalized to a statistical error are presented. Vertical bars are calculated positions of the diffraction peaks corresponding to the atomic and magnetic structures of Ba_2FeMoO_6 (sp. gr. I4/mmm).

versus pressure was interpolated by the Birch-Murnaghan equation [20] (Fig.6):

$$P = \frac{3}{2} \cdot B_0 \cdot (x^{-\frac{7}{3}} - x^{-\frac{5}{3}}) \cdot [1 + \frac{3}{4} \cdot (B_x - 4) \cdot (x^{-\frac{2}{3}} - 1)] \quad (2)$$

where $x = (V/V_0)$ is the relative volume change, V_0 is the unit cell volume at P = 0, B_0 and B' are empirical parameters that have the meaning of the modulus of comprehensive compression in the equilibrium state ($B_0 = -V \cdot (dP/dV)|_{V=V_0}$) and its first order derivative ($B' = dB_0/dP$), respectively. In order to describe the equation of state of Ba_2FeMoO_6 was used a linear dependence corresponding to the Birch-Murnaghan equation for solids with low compressibility:

$$V = (\frac{P}{B_0} - 1) \cdot V_0 \quad (3)$$

Calculated values of B_0 and B' for the equation of state of the Ba_2FeMoO_6 with tetragonal structure are 190 GPa and 4, respectively. The received value of modulus of the all-round compression is reasonable for perovskite structure and well agrees with the results of Ref.[21].

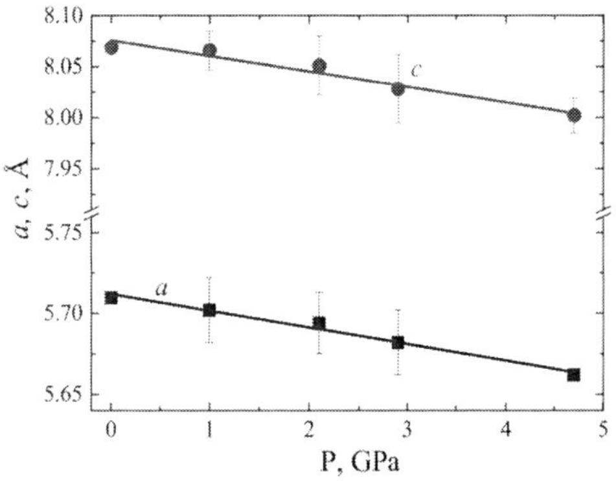

Figure 5. The dependence of crystal lattice parameters of the unit cell Ba_2FeMoO_6 versus external pressure.

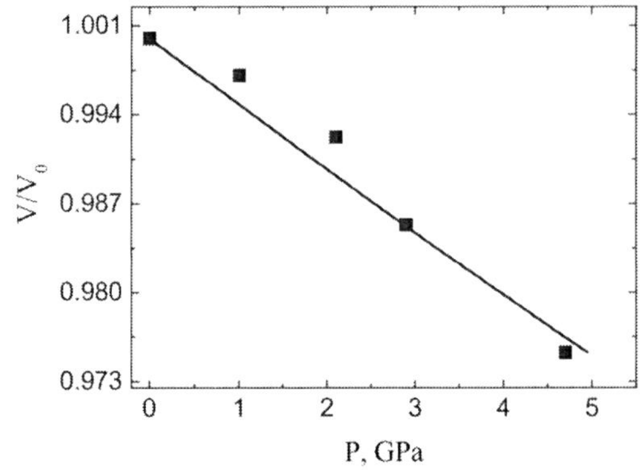

Figure 6. $V/V_0 - P$ is relationship for Ba_2FeMoO_6. The filled squares are experimental data, the solid curve is calculated.

Table 1. Parameters of crystal lattice and atomic coordinates under the influence of ambient pressures at room temperature and figures of merit were refined by the Rietveld method in the framework of the space group: I4/mmm (No. 139). Atoms are in positions: Ba (4d) (*0, 1/2, 1/4*); Fe1 (2a) (*0, 0, 0*); Mo2 (2b) (*0, 0, 1/2*); O1 (4e) (*0, 0, z*); O2 (8h) (*x, x, 0*).

Parameters of unit cell	P, GPa 0	1	2.1	2.9	4.7
a, (Å)	5.7096(3)	5.70(2)	5.69(6)	5.68(4)	5.66(2)
c, (Å)	8.0687(4)	8.07(2)	8.05(9)	8.03(4)	8.00(2)
V, ($Å^3$)	263.03(2)	262(1)	260.9(9)	259(1)	256.5(5)
			O1		
z	0.2601(12)	0.253(22)	0.248(54)	0.258(32)	0.241(36)
			O2		
x	0.2567(0)	0.243(0)	0.246(0)	0.241(0)	0.234(0)
m_{Fe}, μ_B/f.un.	0.38(1)	0.33(16)	0.75(9)	0.594(2)	1.19(1)
R_{wp}, %	14.7	15.3	19.0	22.2	20.2
R_B, %	5.45	5.54	5.69	10.0	8.33
R_{Mag}, %	17.9	20.9	9.36	33.5	10.2
R_F, %	4.36	4.43	4.36	7.09	15.0

The linear parameters of the compressibility of a unit cell of the tetragonal structure were calculated from the formula: $k_i = (1/a_{i0})(da_i/dP)_T$, where the $a_i = (a, c)$ are parameters of the crystal lattice at a pressure of P = 0 and 4.7 GPa. They have the values: $k_a = 0.001774$ (GPa^{-1}) and $k_c = 0.001759$ (GPa^{-1}) at T = 290 K.

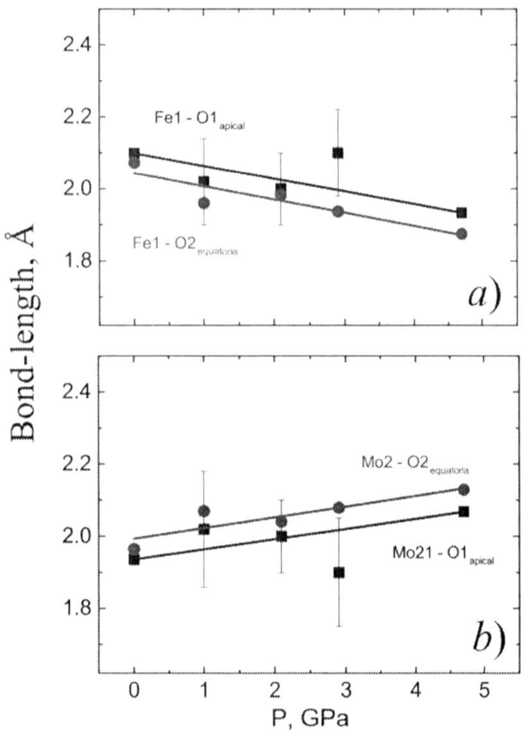

Figure 7. Dependencies of bond-lengths between metal-ligand: *a*) Fe – O and *b*) Mo – O versus pressure.

Table 2. Distances between the metal - oxygen ions in the crystal structure of the double perovskite Ba_2FeMoO_6 determined at various ambient pressures by the Rietveld method within the framework of sp. gr. I4/mmm (No 139).

Ions	Count	P, GPa: 0	1	2.1	2.9	4.7
		Bond-length, Å				
Fe1 - O1	x2	2.09(1)	2.02(16)	2.0(1)	2.1(2)	1.933(4)
Fe1 - O2	x4	2.073(5)	1.96(1)	1.982(2)	1.94(1)	1.875(4)
<Fe1 - O>	x6	2.079	1.98	1.988	1.993	1.894
Mo2 - O1	x2	1.94(1)	2.02(16)	2.0(1)	1.9(2)	2.068(4)
Mo2 - O2	x4	1.965(5)	2.07(1)	2.04(2)	2.08(1)	2.129(4)
<Mo2 - O>	x6	1.957	2.053	2.027	2.02	2.109

IOP Conf. Series: Journal of Physics: Conf. Series **994** (2018) 012014 doi:10.1088/1742-6596/994/1/012014

The effect of external pressure on the magnitude of metal-oxygen bond-length for Ba_2FeMoO_6 in the tetragonal symmetry structure I4/mmm (No 139) is shown in Fig.7 and in Table.2. This crystal structure consists of two types arranged in staggered order asymmetrically distorted oxygen octahedrons FeO_6 and MoO_6 where bond-lengths $Fe/Mo - O1_{apical}$ and $Fe/Mo - O2_{equatoria}$ have different values. Two $O1_{apical}$ coaxial anions located at the vertex of oxygen octahedron FeO_6 along the tetragonal c axis move off iron ion whereas four coplanar $O2_{equatoria}$ ions located in the ab–plane of the octahedron shift to iron ion nearer. The situation in the MoO_6 octahedron is reversed: the coaxial $O1_{apical}$ anions are at a more distance from the Mo ion than four coplanar $O2_{equatoria}$ ions. Bond lengths of $Fe - O1_{apical}$ and $Fe - O2_{equatoria}$, as well as $Mo - O1_{apical}$ and $Mo - O2_{equatoria}$ have close values therefore it is expedient to investigate the pressure dependences of their averaged values: $\langle Fe - O\rangle$ and $\langle Mo - O\rangle$ (Table 2). An increase of the external pressure leads to a reduction of the interionic distances $\langle Fe - O\rangle$ and, conversely, to a slightly increase of the interionic distances $\langle Mo - O\rangle$ (Fig. 7, a and b). Coefficients of the linear compressibility have following values: for the bond $\langle Fe - O\rangle$, $k_{\langle Fe-O\rangle}$ = -0.019 GPa^{-1} and for the $\langle Mo - O\rangle$, $k_{\langle Mo-O\rangle} = 0.017$ GPa^{-1}. The different signs of coefficients of linear compressibility for octahedron FeO_6 and MoO_6 indicate their different behavior under the influence of external pressure.

3.4. Magnetic structure

The electroneutrality in double perovskites can be carried out by two possible valence states of iron and molybdenum ions: 1) Fe^{2+} ($3d^6$) and Mo^{6+} ($4d^0$) or 2) Fe^{3+} ($3d^5$, S = 5/2) and Mo^{5+} ($4d^1$, S = 1/2), respectively. In the first case, the magnetic moment of molybdenum ions is $\mu_{(Mo^{6+})} = 0\mu_B$ therefore the total magnetic moment of this material will be determined by the nominal value of the ordered magnetic moments of ions in iron sublattice $\mu_{(Fe^{2+})} = 4\mu_B$. In the second case, the magnetic structure of the double perovskite should be formed by the ferromagnetic ordering of the magnetic moments of iron ions in FeO_6 octahedrons and by the antiferromagnetic ordering of the magnetic moments of molybdenum ions in MoO_6 octahedrons with nominal values $\mu_{(Fe^{3+})} = 5\mu_B$ and $\mu_{(Mo^{5+})} = 1\mu_B$, accordingly. Therefore, in the fully magnetically ordered state (in the low-temperature region) the total magnetic moment of Ba_2FeMoO_6 should have 4 μ_B/f.un.

The observation of the diffraction spectra of the sample over a wide range of pressures (from 0 to 4.7 GPa) allows one to analyze the change of its magnetic structure. The presence of additional superstructural magnetic reflexes on the neutron diffractograms (Fig. 4) is not observed. This indicates that in the case of the ferrimagnetic structure its wave vector is $\mathbf{k} = [0,0,0]$. The dependencies of magnetic moments of sublattices of Ba_2FeMoO_6 compound versus ambient pressure are shown in Fig. 8 (a) - for ferromagnetic and (b) for antiferromagnetic models. For ferromagnetic model, the value of magnetic moment of Mo sublattice is equal to 0 whereas for Fe sublattice magnetic moment monotonically increases from 0.15 μ_B/f.un. to 1.17 μ_B/f.un. with increasing of ambient pressure from 0 to 4.7 GPa. Magnetic moments of iron sublattice are ordered along easy magnetization axis, which coincide with tetragonal axis. The scheme of the ferromagnetic structure of Ba_2FeMoO_6 is shown in inset of Fig.8, a. In the case of antiferromagnetic structure, magnetic moments of molybdenum sublattice are ordered in opposite direction to iron sublattice. The scheme of the antiferromagnetic structure of Ba_2FeMoO_6 is shown in inset of Fig.8, b. The value of magnetic moment of iron sublattice monotonically increases up to 1.23 μ_B/f.un., whereas of molybdenum sublattice up to 0.15 μ_B/f.un.

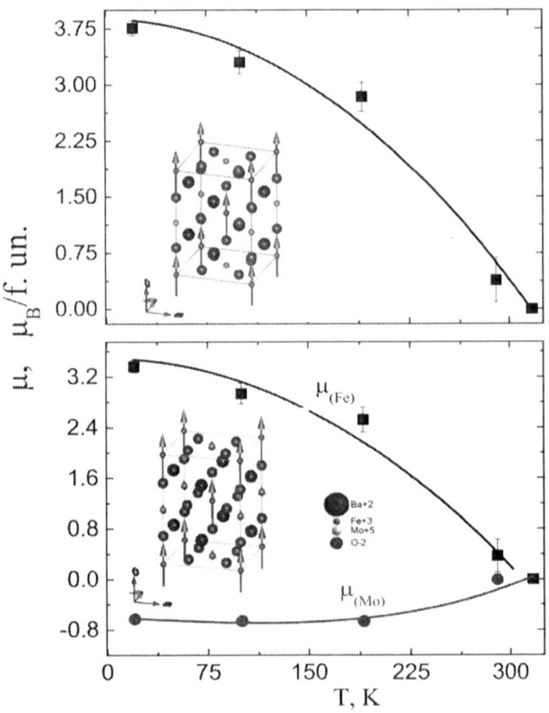

Figure 8. Dependencies of the magnetic moments of iron and molybdenum sublatices of Ba_2FeMoO_6 versus ambient pressure: (a) - for ferromagnetic and (b) - for antiferromagnetic models. In the inset are shown schemes of the magnetic structure of Ba_2FeMoO_6.

4. Conclusions

The behavior of the crystalline and magnetic structures of the double perovskite Ba_2FeMoO_6 has been investigated by the neutron diffraction method in a wide range of ambient pressures (0 - 4.7 GPa). An increase of ambient pressure does not lead to a change the type of crystalline and magnetic structures. It has been determined that a ceramic sample of Ba_2FeMoO_6 prepared by the solid state method under H_2/Ar stream has an ordered tetragonal structure with sp. gr. I4/mmm (No 139) and contains less 10% of anti-site defects. It was shown the influence of ambient pressure to the average interionic metal-ligand distances in the oxygen octahedrons of FeO_6 and MoO_6. The average bond length $\langle Fe - O \rangle$ decreases as the ambient pressure is increased whereas $\langle Mo - O \rangle$ increases. Coefficients of the linear and all-round compressibility were determined for the tetragonal structure of Ba_2FeMoO_6. The average value of the magnetic moment of the iron sublattice increases up to 1.17 μ_B/f.un. monotonically with increasing of ambient pressure in the case of ferromagnetic model and up to 1.23 μ_B/f.un. for antiferromagnetic structure, whereas magnetic moment of molybdenum sublattice increases up to 0.15 μ_B/f.un.

Acknowledgments
The financial supports by the 04-4-1121-2015/2017 project are acknowledged.

References
[1] Patterson F K, Moeller W C and Ward R 1963 *Inorg. Chem.* **2** 196
[2] Colis S, Stoeffler D, Meny C, Fix T, Leuvrey C, Pourroy G, Dinia A and Panissod P 2005 *J. Appl. Phys.* **98** 033905
[3] Park B J, Han H, Kim J, Kim Y J, Kim C S, Lee B W 2004 *J. Magn. and Magn. Mater.* **272-276** 1851
[4] Venimadhav A, Vickers M E and Blamire M G 2004 *Solid State Communications* **130** 631
[5] Huang Y H, Karppinen M, Yamauchi H and Goodenough J B 2006 *Phys. Rev.* B **73** 104408
[6] Ogale A S, Ogale S B, Ramesh R and Venkatesan T 1999 *Appl. Phys. Lett.* **75** 537
[7] Singh V N and Majumdar P 2011 *Europhys. Lett.* **94** 47004
[8] Stoeffler D and Colis S *J. Phys. Cond. Mat.* **17** 6415
[9] Panguluri R P, Xu Sh, Moritomo Y, Solovyev I V and Nadgorny B 2009 *Appl. Phys. Lett.* **94** 012501
[10] Patterson F K, Moeller W C and Ward R 1963 *Inorg. Chem.* **2** 196
[11] Galasso F S 1966 *J. Chem. Phys.* **44** 1672
[12] Kobayashi K-I, Kimura T, Sawada H, Terakura K and Tokura Y 1998 *Nature* (London) **395** 677
[13] Sleight A W and Weiher J F 1972 *J. Phys. Chem. Solid.* **33** 679
[14] Prinz G 1998 *Science* **282** 1660
[15] Kozlenko D P, Savenko B N, Glazkov V P and Somenkov V A 2005 *Neutron News* **16** 13
[16] Piermarini G J, Block S, Barnett J D and Forman R A 1975 *J. Appl. Phys.* **46** 2774
[17] Rodrguez-Carvajal J 1993 *Physica* B **192** 55
[18] Balcells L, Navarro J, Bibes M, Roig A, Martinez B and Fontcuberta J 2001 *Appl. Phys. Lett.* **78** 781
[19] Hemery E 2007 *Magnetic and Transport Studies of Strongly Correlated Perovskite Ceramics* (Victoria University of Wellington, thesis) p 112
[20] Birch F J, 1986 *J. Geophys. Res.* **91** 4949
[21] Yu R C, Zhao P, Li F Y, Liu Z X, Liu J and Jin C Q 2004 *Phys. Rev.* B **69** 214405

Aggregation study in mixture surfactant system TX-100+SDS in heavy water solutions by SANS method

A. Rajewska[1,2], **A. Kh. Islamov**[1], **R. F. Bakeeva**[3]

[1]Joint Institute for Nuclear Research, LNP, 141980 Dubna, Russia
[2]National Centre for Nuclear Research, 05-400 Świerk-Otwock, Poland
[3]Kazan State Technological University, 420015 Kazan, Russia

aldonar@jinr.ru

Abstract. The mixing of amphiphiles in water may lead to the formation of mixed micelles which often present new properties with respect to the pure component solutions [1,2]. The mixture system of classic surfactants SDS (sodium dodecyl sulfate)+TX-100(p-(1,1,3,3-tetramethyl) poly(oxyethylene) (anionic + non-ionic) in heavy water solutions was investigated at temperatures 30°, 50°, 70°C for compositions 1:1, 2:1, 3:1 by the small-angle neutron scattering(SANS) method on spectrometer ('YuMO') at the IBR-2 pulsed neutron source at FLNP, JINR in Dubna (Russia). Measurements have covered Q range from $8x10^{-3}$ to 0.4 $Å^{-1}$. From the measured dependence of the scattered intensity on the scattering angle, we derived the size, shape of micelles, aggregation number at various compositions and temperatures. The size of mixed micelle is a weak function of the mixing ratio between the two components.

1. Introduction

Currently, the structures and properties of mixed solutions formed by two surfactant types, especially by nonionic and ionic surfactants are extensively studied by experimental and theoretical methods, since properties of mixed systems are in many cases better than properties of individual components. Taking into account the fact nonionic surfactants tend to ideal mixing, it can be expected that large deviations from the ideal behavior should appear for mixtures with certain interactions between two surfactant types. Although the synergistic behavior of mixed surfactant systems has long been studied, acquisition of the data on the behavior of various surfactant molecule types in mixed solutions remains an urgent problem. The aim of this work is to study mixed systems of nonionic surfactant TX-100 with anionic surfactant by small angle neutron scattering with the determination of the micelle shape and size. The small angle neutron scattering (SANS) is efficient methods for obtaining information on the structure of mixed micelles and is often used in practice. The critical micelle formation concentration (CMC) was also determined during the study by the surface tension method.

2. Experimental section

Materials

Triton X-100 nonionic classic surfactant (p-(1,1,3,3-tetramethyl) poly(oxyethylene) with CMC= $0.2x10^{-3}$M was obtained as a gift from JINR, Dubna, (Russia). Classic surfactant sodium dodecyl sulfate (CMC = 8.3x10^{-3}M) were purchased from Fluka(99.9 %). These two substances we used without further purification. Heavy water D_2O (99.9%) was purchased in Prikladnaya Chimia" in St.Petersburg (Russia).

2.1. Measurement of CMC - surface tension method

Du Nouy ring method we used for measured the CMC of investigated solutions. The CMC values for TX-100+SDS+D_2O micellar solutions are presented in Table 1.

Table 1. Influence of temperature on CMC for mixed system TX-100+SDS+D_2O for different compositions.

composition of mixture C_{TX-100} / C_{SDS}	$C_{CMC} \cdot 10^{-4}$,M		
	temperature		
	30°C	50°C	70°C
1 : 1	1.00	0,82	0,59
2 : 1	1,50	1,30	1,20
3 : 1	1,60	1,50	1,30

2.2. SANS measurements-small angle neutron scattering (SANS) method

The SANS spectra were measured using a YuMO diffractometer by the time of flight at the IBR-2 pulsed reactor (Joint Institute for Nuclear Research, Dubna, Russia). The specific features of this diffractometer were described in details previously [3]. Neutrons in the range of 0.8–10 A^{-1} were used. Solutions were poured into tightly closed quartz cells (Hellma) 2 mm thick. The temperature within cells (30°, 50°, 70° ± 0.5°C) was maintained constant during measurements using a thermostat. The measured differential scattering cross section was calibrated using a vanadium reference sample. Data pre-processing was performed using the SAS program [4]. For all samples, heavy water (D_2O) was used as a solvent to achieve good contrast conditions. The experimental results were processed using the PCG Software 2.01.02 program (Austria).

3. Results and discussion

Effect of nonionic surfactant TX-100 concentration on structure of micellar solutions in mixed [5-8] system (TX-100+SDS+D_2O) for different compositions and temperatures

The experiment was performed at temperatures of 30°C,50°C and 70°C for three mixed solutions: (i) nonionic surfactant TX-100 (c_1 = 0.022mol/l) in D_2O with anionic surfactant SDS concentrations c =0.022mol/l (ii) nonionic surfactant TX-100(c_2 = 0.044mol/l) in D_2O with anionic surfactant SDS for concentration c =0.022mol/l; (iii) nonionic surfactant TX-100 (c_3 = 0.066mol/l) in D_2O with anionic surfactant SDS for concentration c =0.022mol/l. The effect of nonionic surfactant TX-100 on the structure of micellar solutions of mixed surfactants system TX-100+SDS+D_2O was studied by small angle neutron scattering method as a function of composition and temperature (Figure 1).

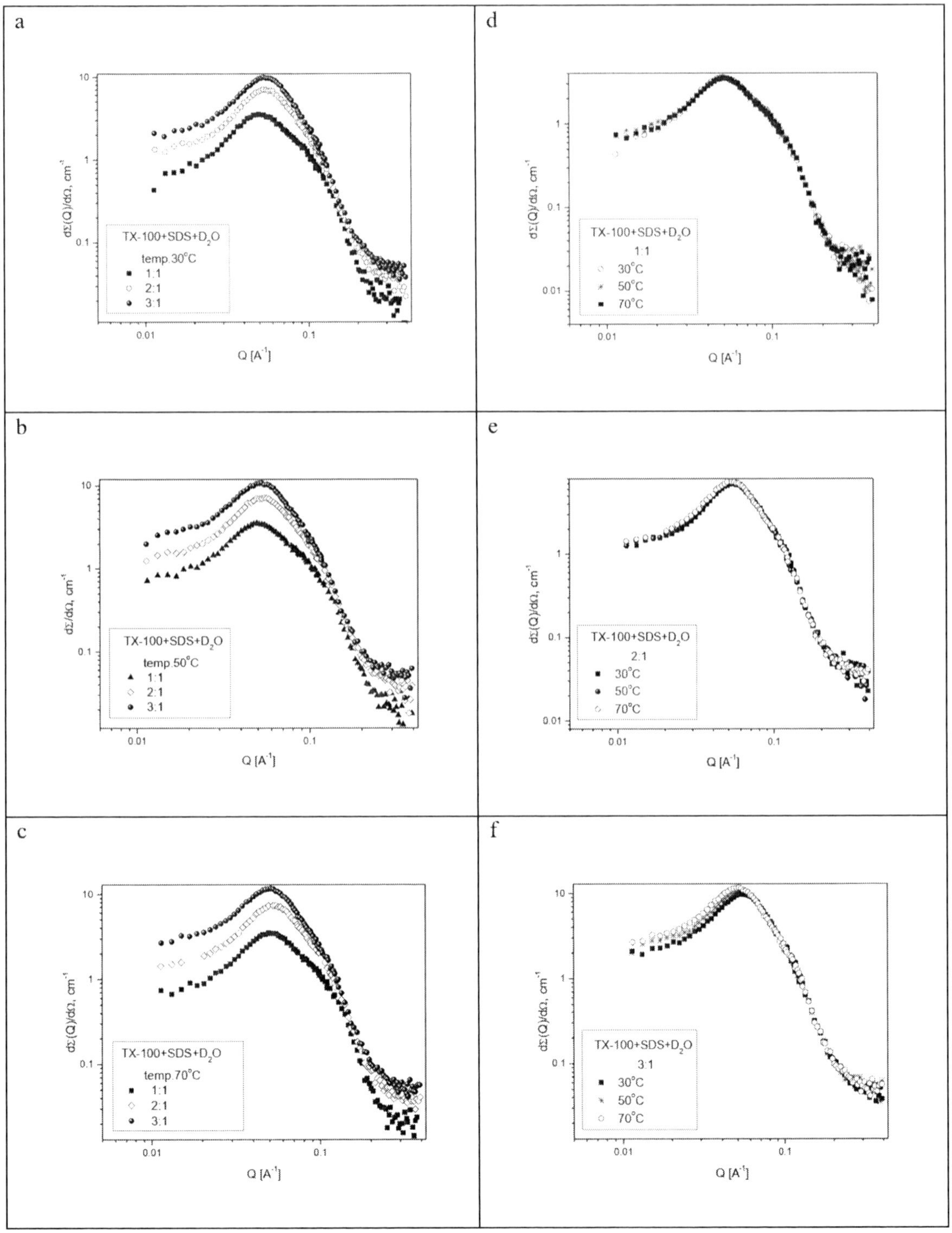

Figure 1. Intensity of neutron scattering vs scattering vector for mixture solution of TX-100+SDS+D_2O at temperatures 30°C, 50°C,70 ° C for three various mix ratios 1:1(a),2:1(b),3:1(c) and for three various mix ratios 1:1(d),2:1(e), 3:1(f) at temperatures 30°C, 50°C,70 ° C.

Table 2. Average distance between micelles (d) and aggregation number (N_{agg}) in micellar solutions for mixture system TX-100+SDS+D_2O

	average distance between micelles (1) aggregation number (2)					
composition	30°C		50°C		70°C	
	1	2	1	2	1	2
1:1	119 Å	26	110 Å	35	120 Å	40
2:1	119 Å	33	111 Å	38	121 Å	38
3:1	125 Å	60	123 Å	58	123 Å	58

The range of measured temperature was 30°C - 70 °C. The higher intensity of neutron scattering is for higher concentration with composition 3:1 but lower intensity is for mix ratio 1:1 at all three constant temperatures (Figure 1 a,b,c). In case of 1:1 composition the position of maximum curves for increases temperatures shifts to smaller Q values at temperature 30°C (Q_{30}=0.049Å^{-1}). The experimental curves for composition 1:1 and 2:1 shown that the position of the maximum intensity is approximately the same at three temperatures 30°C, 50°C and 70 °C (1:1 => Q_{30}= Q_{50}=Q_{70}0.049Å^{-1}; 2:1 => Q_{30}= Q_{50}=Q_{70}=0.054Å^{-1}) but intensity of the scattering neutrons increasing with increase concentration of solutions and mix ratio. For highest temperature70 °C the shift of maximum intensity of neutron scattering was observed for all three mix ratio (Q_{30}= 0.079A^{-1},Q_{50}=0.052A^{-1},Q_{70}=0.050A^{-1}). From these results the conclusion is that with increasing ratio between concentration of non-ionic surfactant TX-100 and anionic SDS from 1:1 to 3:1 (increasing amount of non-ionic surfactant) the system changes properties very slowly from non- ionic like to ionic like for the highest mix ratio with increase temperature.
The intensity of neutron scattering for mix ratio 1:1 and 2:1 is different but we not observed changes intensity at different temperature (Figure 1 e, f, g). The intensity of scattering neutrons for the mix ratio 3:1 is highest than for the mix ratios 1:1 and 2:1. The maximum of the scattering neutrons intensity increases with increase of mix ratios. Shift of the maximum of intensity to the lower value of scattering vector Q was observed for concentration 3:1 (Q_{30} = 0.079A^{-1}; Q_{50}=0.052A^{-1};Q_{70}=0.050A^{-1}) with increase of temperature. For the 3:1 mix ratio the intensity of neutron scattering for all three curves and lower scattering vector Q increases with increase temperature. The experimental results for CMC values for all investigated solutions shown that CMC changes with the composition of solutions. In mixed system solutions are mixed micelles. In Table 2 are the values of the average distance between micelles d and aggregation number N_{agg}. The micelles are rather small and spherical in all investigated solutions.

Conclusions

The preliminary results obtained in this study can be summarized as follows. For all the studied aqueous solutions with constant amount of anionic surfactant SDS+D2O with additions of non-ionic surfactant TX-100, the existence of micelles was detected. It was found that spherical micelles were formed in mixed system TX-100+SDS+D2O in all studied solutions. The aggregation number for the solutions with compositions 1:1 and 2:1 is smaller at temperature 30°C than at 50°C but highest at 70°C. For mix ratio 3:1 the aggregation number was approximately constant for all temperatures (30°C,50°C,70°C). The aggregation number for mixed system with the highest concentration (3:1) increases with increase of mix ratio. The average distance between micelles in investigated micellar mixed systems at three different concentrations (at constant anionic SDS surfactant concentration) was constant with increasing concentration of the non-ionic TX-100 surfactant. It was observed shift of the maximum intensity for these experimental curves. The intensity of scattering neutrons as a function temperature was constant for all temperatures for three different mix ratios. There are only preliminary experimentally results and the next paper is ready for these mixed micellar solutions.

IOP Conf. Series: Journal of Physics: Conf. Series **994** (2018) 012015 doi:10.1088/1742-6596/994/1/012015

References

[1] Asakawa T, Yohten K, Miyagishi S, Nishida M 1985 *Langmuir* **1** 347
[2] Brasher L L, Kaler E W 1996 *Langmuir* **12** 6270
[3] Ostanievich Yu M 1988 *Makromol. Chem. Macromol. Symp.* **15** 91
[4] Solov'ev A G, Solov'eva T N, Stadnik A V, Islamov A Kh, Kuklin A I 2003 *Soobshch. Ob'edin. Inst. Yad. Issled. Dubna* **86** 10
[5] Almgren M, Hansson P, Wang K 1996 *Langmuir* **12** 3855
[6] Herrington K L, Kaler E W, Miller D D, Zasadzinski J A, Chruvolu S 1993 *J. Phys. Chem.* **97** 13792
[7] Rajewska A, Mędrzycka K, Hallmann E 2014 *Physics of the Solid State* **56** 125
[8] Rajewska A, Mędrzycka K, Hallmann E 2007 *Kristallografiya* **52** 811

High-throughput SANS experiment on two-detector system of YuMO spectrometer

A I Kuklin[1], A I Ivankov[1,2], D V Soloviov[1,3], A V Rogachev[1], Yu S Kovalev[1], A G Soloviev[1], A Kh Islamov[1], M Balasoiu[1,4], A V Vlasov[5], S A Kutuzov[1], A P Sirotin[1], A S Kirilov[1], V V Skoi[1,6], M I Rulev[1] and V I Gordeliy[7]

[1]Joint Institute for Nuclear Research, Dubna, Russia
[2] Institute for safety problems of nuclear power plants NAS of Ukraine, Kyiv, Ukraine
[3]Taras Shevchenko National University of Kyiv
[4]Horia Hulubei National Institute for Physics and Nuclear Engineering, Bucharest, Romania
[5]Institute of Crystallography, RWTH Aachen University, Aachen, Germany
[6]Lomonosov Moscow State University, Moscow, Russia
[7]Institut de Biologie Structurale, J.-P. Ebel, Université Grenoble Alpes-CEA-CNRS, F-38000 Grenoble, France

E-mail: kuklin@nf.jinr.ru

Abstract. Using a multidetector system on the YuMO spectrometer allows shortening the time of measurements. The quantitative comparison of the measurement time using one and two-detector mode is done. The time range for experiments was from several minutes up to 12 hours. It was shown that two-detector system shortens more than twice the time of the measurement. While making a structural investigation using advanced software the two-detector system allows to treat the data at a qualitatively new level. An example illustrating the features of the channels choice and measurement time on the spectrometer was shown. The results of this paper could be used when planning the experiments on the YuMO spectrometer, for modernization of the installation and for equipment using time-of-flight method.

1. Introduction

For the most commonly used scientific instruments the time of measurements is defined by experimenter. Occurrence of the spectrometers on reactors has changed this paradigm. The operation time of the spectrometer became strongly dependent on the reactor operation time. The experiment went into industrial scale. The increasing interest in nanotechnologies and consistent structural problems [1, 2] has raised the busyness of reactor instruments. The difficulty of shortening the time of measurements without the loss of data quality has occurred especially for time-of-flight facility. The problem could be solved via full automatization of the spectrometer [3, 4] and eliminating the experimenter-made errors.

The system that uses two detectors simultaneously [5-7] was designed and mounted on the small angle spectrometer YuMO on the reactor IBR-2. It allowed significantly shortening the time of experiment. The system was also extremely important especially because modern data treatment software requires wide Q-range [8-12] in contrary to the earlier periods of small angle neutron scattering development [13]. The problem of experimenter at the moment is to obtain data in wide Q-range with satisfying statistical error, which is determined by the time of measurement. The measurement time depends on the scattering cross-section at the other similar conditions, first of all, the Q-range.

In the work [6] the scheme of the experiment, calibration methods for samples that have low and high scattering, optimization of sizes and distances for vanadium standards are described. The idea of shortening the time of experiment could arise after two decades of two-detector system exploitation and several times increase of scientific publications made with the help of the instrument [7].

A new word in modernization of the YuMO spectrometer is a design of the second unique position sensitive detector (PSD) in the framework of two-detector project. As soon as the question about duration of the exposure on the small angle neutron spectrometer on the reactor IBR-2 was still unclear, it will be described in details in this work.

2. Time of measurement factors

The time of measurement significantly depends on the cross-section scattering of the sample. The available intensity range is determined by the statistical error that directly depends on the measurement time. Another factor is the wideness of the Q-range that is necessary for experiment. If the value is bigger than the dynamic range of the spectrometer (q_{max} / q_{min}) the experimenter should make additional measurements. For stationary reactors producing monochromatic irradiation the dynamic range is determined by the size of the detector. For pulsed neutron sources (accelerator, IBR-2) the terms such as Q-range and dynamic Q-range are quite relative. The limitation of the YuMO spectrometer for samples that have a low scattering is nearly 0.05 - 0.1 cm^{-1}. In this case the Q-range in Guinier approximation area is about 0.02 A^{-1} and in the area of Porod approximation is up to 0.2 A^{-1}. The dynamic range in this case is only 10. For the samples that have high scattering the Q-range is 0.006 – 0.6 A^{-1}, and the dynamic range is more than 100. Consequently, the Q-range, dynamic Q-range and time of measurement significantly depend on the sample. It is worth to highlight that the samples with low scattering when it seems that using one detector satisfies the dynamic Q-range, the time of measurements could also be shortened twice if using the second detector in optimal position. This happens due to the using of the time-of-flight method and the possibility of moving the detectors so that the Q-range will be approximately similar for both detectors. In addition, experimenter often makes test measurements with samples that could have specific behavior in different points of Q-range. Such experiments commonly last several minutes and experimenter does not move the detector. In case of one-detector system the effect could be lost. Consequently the using of two-detector system the data quality is much higher.

The smoothing technique [14] for small angle data could also shorten the time of measurement except "tails" effect. This could be achieved only by using of two-detector system and advanced data treatment methods.

3. The comparison of the time of measurements: one detector at two positions *vs* two detectors simultaneously

The question about the comparison of the time of measurements (one detector at two positions and two detectors simultaneously) could be reduced to the question of the time of measurements in the nearest position (big values of Q) and the further position (small values of Q) at the optimal collimation for each position in case of one detector. These times of measurements are almost similar. So the low scattering cross-section at big values of Q (the nearest position) and the consequent necessity of long exposure is compensated the small number of neutrons and consequently almost the same exposure time at low values of Q (the further position). It is worth to point out that the measurement at nearest position (for shortening the time of the experiment) implies information loss for monodisperse samples (dendrimers, proteins, tracking membranes) and samples that have a structural factor (lipid membranes, polymers, etc.). Two-detector system has no such disadvantage. The need of high resolution at the nearest position for the wide spectrum of samples (alconates [15], proteins [16-19], dendrimers [20-25], nanotubes [26], lipids [27-29]) leads to conducting the experiments at the nearest position of the well-collimated detector. Consequently the time of measurements range on the spectrometer YuMO is from 3 min [30, 31] up to 12h [32, 33].

On the spectrometer YuMO there is a specific procedure of normalization [34, 35]. For two-detector system this procedure has some corrections [6]. The same reasoning is true for the time of measurements with and without vanadium standards [36]. There is also the ratio

$$t_{s+v} / t_s = (I_s / I_{s+v})^{1/2} \qquad (1)$$

where t_{s+v} is the time of measurement of the sample simultaneously with vanadium standard into the beam, t_s is the time of the measurement of the sample, I_s – scattering intensity of the sample, I_{s+v} – scattering intensity of the sample and vanadium standard in the beam simultaneously. The ratio could be defined by the minimum scattering of the vanadium standard. It is important that the longer the measurement of the background the wider the dynamic range.

Here is the qualitative evaluation, because the quantitative calculation requires exact values of Q-range and depends on the sample. The efficiency of detector grows versus wavelength and the transmission decreases almost at the same order, so these coefficients cancel each other. The ratio of fluxes for different wavelengths [37] is nearly 10^{-4} and the ratio of differential cross-sections is some units of 10^3 and the ratio of solid angles is approximately 1/6. So, taking in account the square root from the formula (1) the times of measurements is the same order values. This is only a qualitative evaluation. For every sample the ratio of the times of measurements should be calculated individually. In practice the experimenter has an adequate statistical error taking into account the real operation time and consequently correcting big or low values of Q so that the times of measurements at small angle and wide angle area of Q are equal in the case of two-detector system.

The SAS package [35, 38-39] is used to process the spectra measured on YuMO spectrometer.
And with modernizing the spectrometer SAS has been developing in parallel. It has preserved the fundamentals of TOF data treatment, but its architecture was refactored significantly. In general, processing with SAS includes these main stages:

- normalization of the time spectra over the scattering from the vanadium standard;
- conversion time spectra into the space of neutron momentum transfer.

First stage is performed independently for each detector ring. And on the second one the resulting spectra are averaged over the set of rings.

Modern SAS does not operate with "OLD" and "NEW" detectors separately, as before, when spectra for them have to be sewed. Instead, it operates with the whole set of rings for both detectors and produces the final spectrum in a natural way [35]. Moreover, this approach improves the statistics for the middle region of the averaged spectra. This region corresponds to the boundary between "OLD" and "NEW" detectors, and when averaging them separately boundary effects made it difficult to sew the spectra.

In the Fig. 1 there are SANS curves for carbon powders smoothed (for clarity) as described in [14]. Guinier approximation area of the curve has Q-shift due to decreasing of time channels range for obtaining the scattering picture at «true» intensity scale. Otherwise the scattering intensity will be in arbitrary units so the slopes of the scattering curves (Fig. 1) and the discrepant parts of the curves in the Guinier approximation region are not represent the true scattering picture. The same is also true for the right border. However, there the effect is not so prominent because of the small scattering cross-section.

The fact that SANS intensities for 60, 30 (overlap fully visually) and 10 min of measurements are almost similar means that for samples with high scattering the time of measurement 10 min is enough. However, there are some discrepancies near the right border.

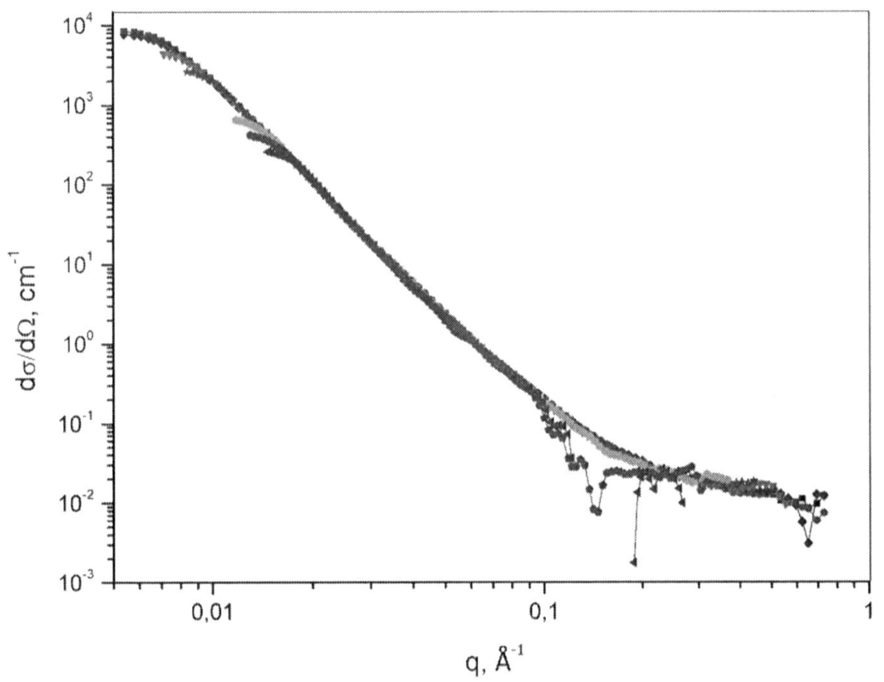

Figure 1. Powder carbon SANS curves for different measurement time. For some curves the selection of TOF channels was managed. From left curves to right: time of measurement 60 min (55-450 time-of-flight channels), 30 min (55-450 time-of-flight channels), 10 min (55-450 time-of-flight channels), 5 min (65-350 time-of-flight channels), 2.5 min (75-300 time-of-flight channels), 70 second (85-220 time-of-flight channels), 30 second (95-200 time-of-flight channels). First two curves (60 min and 30 min) overlap fully visually.

The optimal Q-range has especially to be taken into account when measuring samples with «fractal» properties. The fractal structures with determined dimension have a linear behavior of scattering curve in log-log scale [40-45]. Boundary conditions determine the size-range where the fractal structure could be observed. That is why the optimal Q-range is crucial in this case.

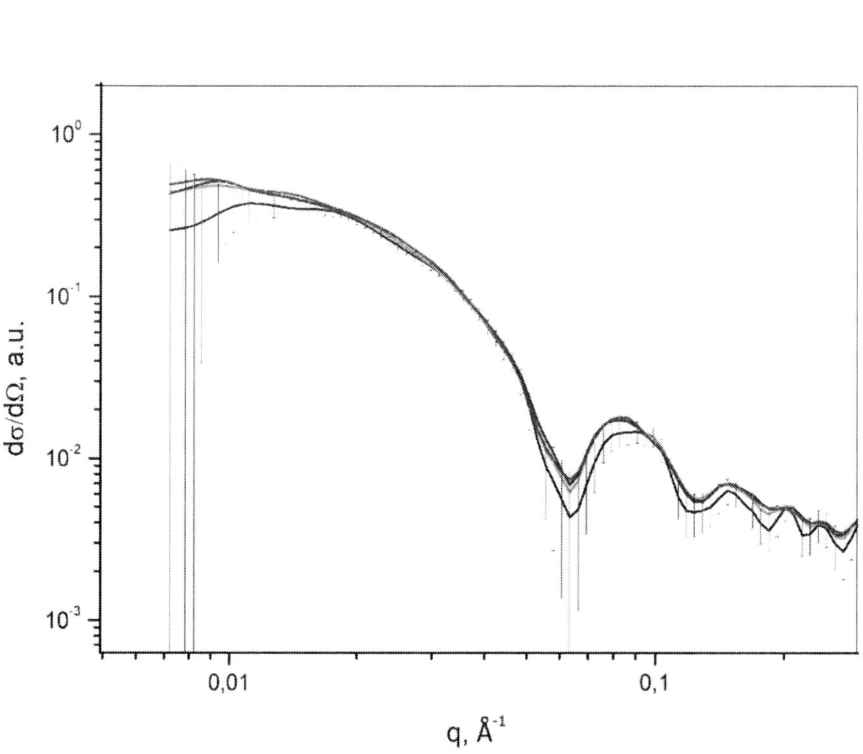

Figure 2. Apoferritin solution with 7 mg/ml concentration in capillary (2 mm light path) SANS curves for different measurement times: 80 min (red curve), 60 min (blue curve), 40 min (yellow curve) and 20 min (black curve). All curves are smoothed for obvious vision by procedure, described in [14].

In Fig.2 shown example of curves demonstrates good possibility of measurement in capillary 2 mm light path.

4. Additional time loss during experiments with one-detector system

Another difference between one- and two-detector system modes of data collection on the spectrometer YuMO is the following.

For example if there is an experiment on the spectrometer YuMO that includes measurements of 25 samples with 4 different temperatures each (in 30 degrees centigrade range) using one-detector system in two different positions or two-detector system. Additional time loss connected with detector moving 10 meters is about 0.25h. Nearly the same time loss is for changing the samples. To change the temperature – minimum 30 min and for its stabilization – 1.5h. The resulting time loss is nearly 3h. If the time of measurement is 20 min for each sample (high scattering samples) then there is 65% of time loss. If there are low scattering samples (the time of measurement is 60 min for each) the loss is about 20%. If the time of measurement of buffer simultaneously with temperature changing is taken into account (despite it is not always correct) the ratio of the time loss will be lower, but nevertheless significant part of the operation time.

Besides, the nearest to the sample detector decreases the background for the second detector, that increases the signal/noise ratio and results in shortening the time of measurement. The quantitative evaluation of this effect is not simple enough and is out of the scope of this paper.

Concerning kinetic experiments and unique samples with short lifetime, clustering effects (especially biological or colloids) the only way of obtaining structural information in wide dynamic range is using two-detector system. This is also works for investigating samples at temperature and/or pressure changes and phase transitions.

5. Conclusion

The dynamic range and Q-range are significantly dependent on sample. Consequently, as the time of measurement for most of the samples in near position is practically the same as in far position, the whole time of measurement shortens twice. While measuring samples with low scattering intensity and consequently narrow Q-range the second detector could twice the statistical data collection that is equal to shortening twice the time of measurement if optimal positions for the both detectors have been setup.

When using only one detector there is an additional loss of the time of measurement (from 20 to 60%) connected with detector movement, sample and temperature changing. It means that the time shortening for two-detector system in comparison with one-detector system is even more than twice for most of the experiments. Expected ratio of the times of measurement for other small angle spectrometers is nearly the same if the sizes of detectors are equal.

The results of the paper are proved by almost two decade experience of two-detector system exploitation on the YuMO spectrometer.

6. Acknowledgement

The work was supported by the Federal Ministry of Education and Science (BMBF), Germany, grant No. 03DU03G2, grants from Plenipotentiary Representatives of Romania, Slovakia, Czech Republic and Poland.

References

[1] Roco M C, Williams R S and Alivisatos P 2000 Nanotechnology research directions. IWGN workshop report. Vision for nanotechnology R&D in the next decade (Springer Science & Business Media)
[2] Sergeev G B 2006 Nanochemnistry (Elsevier)
[3] Kuklin A I *et al.* 2004 Automation and environment of a sample of the modernized installation YuMO *Preprint of the Joint Institute for Nuclear Research* **13-2004-77**
[4] Kuklin A I *et al.* 2011, *J. Phys.: Conf. Ser.* **291** 012013
[5] Kuklin A I, Islamov A Kh and Gordeliy V I *Neutron News* **16 3** 16
[6] Kuklin A I *et al.* 2006 *Journal of Surface Investigations, X-ray, Synchrotron and Neutron Techniques* **6** 74 (in Russian)
[7] Kuklin A I, Rogachev A V, Soloviov D V, Ivankov O I, Kovalev Yu S, Utrobin P K, Kutuzov S A, Soloviev A G, Rulev M I and Gordeliy V I 2017 *J. Phys.: Conf. Ser.* **848 1** 012010
[8] Svergun D I 1992 *J. Appl. Crystallogr.* **25** 495
[9] Svergun D I, Barberato C and Koch M H J 1995 *J. Appl. Crystallogr.* **28** 768
[10] Svergun D I 1999 *Biophysical Journal* **76** June 2879
[11] Konarev P V, Petoukhov M V and Svergun D I 2001 *J. Appl. Crystallogr.* **34** 527
[12] Petoukhov M V, Nigel A, Eady J, Brown K A and Svergun D I 2002 *Biophysical Journal* **83** December 3113
[13] Feigin L A and Svergun D I 1987 Structure analysis by small-angle X-ray and neutron scattering (New York: Plenum Press) 335
[14] Soloviev A G, Litvinenko E I, Ososkov G A, Islamov A Kh and Kuklin A I 2003 *Nuclear Inst. and Methods in Physics Research, A.* **502/2-3** 498
[15] Nyam-Osor M, Soloviov D V, Kovalev Yu S, Zhigunov A, Rogachev A V, Ivankov O I, Erhan R V and Kuklin A I 2012 *J. Phys.: Conf. Ser.* **351** 012024
[16] Lebedev D V, Baitin D M, Islamov A Kh, Kuklin A I, Shalguev V Kh, Lazov V A and Isaev-Ivanov V V 2003 *FEBS Letters* **537** 182
[17] Murugova T N *et al.* 2015 *Journal of optoelectronics and advanced materials* **17** 9-10 1397
[18] Kuklin A I, Murugova T N, Ivankov O I, Rogachev A V, Soloviov D V, Kovalev Yu S, Ishchenko

A V, Zhigunov A, Kurkin T S and Gordeliy V I 2012 *J. Phys.: Conf. Ser.* **351** 012009
[19] Feldman T B, Ivankov O I, Murugova T N, Kuklin A I, Shelyakin P V, Yakovleva M A and Ostrovsky M A 2015 *Doklady Biochemistry and Biophysics* (Pleiades Publishing) **465** **1** 420
[20] Kuklin A I *et al.* 2003 *J. Appl.Cryst.* **36** 679
[21] Ozerin A N, Muzafarov A M, Kuklin A I, Islamov A Kh, Gordelyi V I, Ignat'eva G M, Myakushev V D, Ozerina L A and Tatarinova E A 2004 *Doklady Chemistry* **395, Part 2** 59 (translated from 2004 *Doklady Akademii Nauk* **395** **No. 4** 487)
[22] Ozerin A N, Svergun D I, Volkov V V, Kuklin A I, Gordeliy V I, Islamov A Kh, Ozerina L A and Zavorotnyuk D S 2005 *J. Appl. Cryst.* **38** 996
[23] Kuklin A I *et al.* 2011 *Rom. Journ. Phys.* **56** **1–2** 134
[24] Rogachev A V, Kuklin A I, Cherny A Yu, Ozerin A N, Muzafarov A M, Tatarinova E A and Gordeliy V 2010 *Physics of the Solid State* **52** 5 1045
[25] Rogachev A V, Cherny A Yu, Ozerin A N, Muzafarov A M, Tatarinova E A, Islamov A Kh, Gordeliy V I and Kuklin A I 2008 *J. Phys.: Conf. Ser.* **129** 012041
[26] Bulavin L A, Savenko V S, Lebovka N I, Kuklin A I, Soloyiov D V and Ivankov O I 2013 *Nuclear Physics and Atomic Energy* (Institute for Nuclear Research of NAS Ukraine) **14** **4** 372
[27] Soloviov D, Zabashta Y, Bulavin L, Olexandr I, Gordeliy V and Kuklin A 2014 *Investigations in Macromolecular Symposia* **335** 1 58
[28] Bulavin L A, Soloviov D V, Gordeliy V I, Svechnikova O S, Krasnikova A O, Kasian N A and Lisetski L N 2015 *Phase transitions* **88(6)** 582
[29] Bulavin L A, Solov`ev D V, Gorshkova Yu E, Zhigunov A M, Ivan'kov O I, Gordelii V I, Kuklin A I 2012 *Ukr. J. Phys.* **57** **6** 623
[30] Efremov R, Shiryaeva G, Bueldt G, Islamov A, Kuklin A, Yaguzhinsky L, Fragneto-Cusani G and Gordeliy V 2005 *Journal of Crystal Growth* **275** e1453.
[31] Uhríková D, Teixeira J, Hubčík L, Búcsi A, Kondela T, Murugova T and Ivankov O I 2017 *J. Phys.: Conf. Ser.* **848** **1** 012007
[32] Kuklin A I, Ostanevich Yu M, Cser L, Grosz T and Jancso G 1990-1991 *Annual report LNP* 163
[33] Cser L, Farago B, Grosz T, Jancso G and Ostanevich Yu M 1992 *Physica B* **180&181** 848
[34] Ostanevich Yu M 1988 *J. Makromol. Chem. macromol.Symp.* **15** 91
[35] Soloviev A G, Solovjeva T M, Ivankov O I, Soloviov D V, Rogachev AV and Kuklin A I 2017 *J. Phys.: Conf. Ser.* **848,1** 012020.
[36] Bezzabotnov V Yu and Ostanevich Yu M 1989 *Physica B* **156&157** 595
[37] Kuklin A I, Rogov A D, Gorshkova Yu E, Utrobin P K, Kovalev Yu S, Rogachev A V, Ivankov O I, Kutuzov S A, Soloviov D V and Gordeliy V I 2011 *Physics of Particles and Nuclei Letters* **8** **2** 119
[38] Soloviev A G, Solovieva T M, Stadnik A V, Islamov A Kh and Kuklin A I 2003 *Communication of JINR* **P10-2003-86**
[39] https://wwwinfo.jinr.ru/programs/jinrlib/sas/indexe.html
[40] Cherny A Y, Anitas E M, Osipov V A and Kuklin A I 2011 *Physical Review E.* **84(3)** 036203
[41] Cherny A Y, Anitas E M, Osipov V A and Kuklin A I 2014 *Journal of Applied Crystallography* **47(1)** 198
[42] Cherny A Y *et al. J. Synch. Investig* **4: 903** 90
[43] Cherny A Y, Anitas E M, Kuklin A I, Balasoiu M and Osipov V A 2010 *Journal of Applied Crystallography* **43(4)** 790
[44] Cherny A Y, Anitas E M, Osipov V A and Kuklin A I 2017 *Physical Chemistry Chemical Physics* **19(3)** 2261
[45] Cherny A Y, Anitas E M, Osipov V A and Kuklin A I 2017 *Journal of Applied Crystallography* **50(3)** 919

IOP Conf. Series: Journal of Physics: Conf. Series **994** (2018) 012017 doi:10.1088/1742-6596/994/1/012017

Ambiguities and completeness of SAS data analysis: investigations of apoferritin by SAXS/SANS EID and SEC-SAXS methods

D V Zabelskii[1,2,α], A V Vlasov[1,3,α], Yu L Ryzhykau[1], T N Murugova[1,4], M Brennich[5], D V Soloviov[1,4,6], O I Ivankov[1,4,7], V I Borshchevskiy[1], A V Mishin[1], A V Rogachev[1,4], A Round[8], N A Dencher[1,9], G Büldt[1], V I Gordeliy[1,2,10] and A I Kuklin[1,4]*

[1]Research center for molecular mechanisms of aging and age-related diseases, Moscow Institute of Physics and Technology, Dolgoprudny, Russian Federation
[2]Institut de Biologie Structurale, J.-P. Ebel, Université Grenoble Alpes-CEA-CNRS, Grenoble, F-38000, France
[3]Institute of Crystallography, RWTH Aachen University, Germany
[4]Joint Institute for Nuclear Research, Dubna, Russian Federation
[5]European Molecular Biology Laboratory, Grenoble Outstation, CS 90181, 71 Avenue des Martyrs, 38042, Grenoble Cedex 9, CS 90181, France
[6]Shevchenko National University of Kyiv, Kyiv, Ukraine
[7]Institute for safety problems of nuclear power plants NAS of Ukraine, Kyiv, Ukraine
[8]European X-Ray Free-Electron Laser Facility GmbH at Amtsgericht Hamburg, HRB 111165
[9]CSI Organic Chemistry and Biochemistry, Technische Universität Darmstadt, Darmstadt, D-64287, Germany
[10]Institute of Complex Systems: Structural Biochemistry (ICS-6), Research Centre Jülich, Jülich, 52425, Germany,
[α]-equal contribution
kuklin@nf.jinr.ru

Abstract. The method of small angle scattering (SAS) is widely used in the field of biophysical research of proteins in aqueous solutions. Obtaining low-resolution structure of proteins is still a highly valuable method despite the advances in high-resolution methods such as X-ray diffraction, cryo-EM etc. SAS offers the unique possibility to obtain structural information under conditions close to those of functional assays, i.e. in solution, without different additives, in the mg/mL concentration range. SAS method has a long history, but there are still many uncertainties related to data treatment. We compared 1D SAS profiles of apoferritin obtained by X-ray diffraction (XRD) and SAS methods. It is shown that SAS curves for X-ray diffraction crystallographic structure of apoferritin differ more significantly than it might be expected due to the resolution of the SAS instrument. Extrapolation to infinite dilution (EID) method does not sufficiently exclude dimerization and oligomerization effects and therefore could not guarantee total absence of dimers account in the final SAS curve. In this study, we show that EID SAXS, EID SANS and SEC-SAXS methods give complementary results and when they are used all together, it allows obtaining the most accurate results and high confidence from SAS data analysis of proteins.

1. Introduction

Small-angle scattering (SAS) is a powerful technique for investigating large-scale structures at scales from 10Å to 10^4Å and is widely used in biology and material science [1–20] especially for investigations of complexes [21–26] and large structures [27–29], Consequently, the question of quantitative characterization of these structures becomes extremely important. In structural biology, small-angle X-ray and neutron scattering (SAXS and SANS) are mostly used to determine integral structural parameters of biomolecules such as radius of gyration, scattering volume or aggregation parameters. SAXS and SANS allow one to recover three-dimensional shape from one-dimensional scattering pattern by dummy atom modelling [30]. These approaches require high purity and monodispersity of samples in order to exclude clustering effects by using standard sample preparation protocols and data processing methods such as extrapolation to infinite dilution (EID). The EID method uses extrapolation of I(q) curves for different concentrations of the protein to zero concentration for creating idealized scattering curve [31]. The method works well for excluding of structure factor effects. In case there are dimers or aggregates and the dependency on concentration is sophisticated, we suggest using small angle X-ray scattering method combined with size-exclusion gel-chromatography (SEC-SAXS) method. SEC-SAXS uses on-line high-pressure liquid chromatography (HPLC) system coupled with SAXS instrument that enables direct SAXS measurements of chromatographic sample fractions. Originally this method was developed to allow measurements of unstable and aggregating samples, but it can also be used for stable proteins [32–39].

In this investigation, we show dummy atom models reconstructed from the scattering curves. We use apoferritin due to its simplicity, stability and highly spherical shape [40–42]. Apoferritin is an iron depleted form of ferritin which is well known for its important role as main iron storage [43–45]. In accordance with X-ray crystallography data, apoferritin consists of 24 subunits forming a quasi-spherical core shell with a cavity inside [46] and several intersubunit channels that communicate between the interior of the molecule and external environment. Apoferritin may encapsulate small molecules, making it an interesting system for drug delivery [41,47–49]. Moreover, ferritin could be a candidate for a marker of cancer and/or biological age [50] or has a biotechnical implementations like some other proteins such as rhodopsins etc. [51,52]. However, crystallographic structure of the protein state can be influenced by crystallization conditions and crystal packing. Usually authors do not distinguish the influence of the structure factor effects and the effects of oligomerization, taking into account the quadratic dependence of the scattering intensity on the volume of the object under study. Therefore, SAS is an important approach that could be used to determine structural parameters of apoferritin under native conditions.

Apoferritin 3D dummy atom modelling was already done in both small angle X-ray (SAXS) and neutron (SANS) scattering experiments, using EID method to exclude clustering effects [53].

However, the EID method can be insufficient to obtain an idealized monomer SAXS curve with confidence. The quality of SAXS data throughout the full Q-range plays crucial role in data processing and may seriously influence final models and conclusions if not treated correctly.

The aim of this study is a comparative analysis of structural parameters of proteins in solution and in crystal using as an example the protein apoferritin. This requires the greatest possible resolution and elimination of the structure and oligomeric factors. As it was mentioned above, different methods provide different results, which might also depend on the oligomeric factor and concentration of the protein.

2. Materials and Methods

2.1. SEC-SAXS experimental setup and sample description

Horse spleen apoferritin was purchased from Sigma-Aldrich Chemie GmbH (Product number A3660). Storage buffer (50% glycerol and 0.075 M NaCl) was replaced with buffer (200 mM NaCl and 20 mM

Tris, pH=7.5) by dialysis in 1L of the buffer. After dialysis, the protein was applied to Superdex 200 (S200 GL 10/300, GE Healthcare) size-exclusion column, which was connected to on-line HPLC system (LC-20ADXR, Shimadzu) attached directly to the sample-inlet valve of the BM29 sample capillary [39,54]. 500 μL of 8.8 mg/mL protein solution were injected onto the column (Superdex 200 10/300 GL, GE Healthcare). Buffers were degassed and the column was equilibrated with 1.5 column volumes of buffer. A flow rate of 0.5 ml/min at room temperature was used and the baseline was monitored. All data from the run were collected at a wavelength of 0.99 Å using a sample-to-detector distance (PILATUS 1M, DECTRIS) of 2.876 meters corresponding to an effective Q range of 0.008–0.45$Å^{-1}$. Approximately 1200 frames (1 frame per 2s) were collected per 40 min sample run and initial data processing was performed automatically using the EDNA pipeline [47,55], generating radially integrated, calibrated and normalized one-dimensional profiles for each frame. All frames were compared with the initial frame and matching frames were merged to create the reference buffer automatically by the EDNA software.

Frames with a consistent Rg from the peak scattering intensity were manually merged to get a single averaged frame corresponding to the scattering of SEC purified monomer fraction. Specifically, 20 frames corresponding to the monomer peak were merged and used for all further data processing and model fitting.

2.2. SEC purification

Horse spleen apoferritin was purified as a soluble protein by SEC-SAXS experimental setup. In Fig.1 two partially resolved peaks represent different fractions of apoferritin in solution. In SEC method fractions with higher molecular weight elutes first, therefore peak fraction "B" corresponds to oligomer fraction that is in a good agreement with Rg and I(0) data. Peak fraction "A" is the apoferritin monomer fraction with high purity and monodispersity that is controlled by Rg data. Frames used for creating the idealized SAXS curve correspond the peak monomer fractions with the highest I(0) and absorbance (range limited by the vertical dash lines in Fig. 1). However, only 20 peak frames were taken for merging due to possible concentration effects that may affect the final curve. Difference in effective concentrations and gyration radii between frames used for merging is less than 10%. We note that more frames can be used for averaging, although, this will give insufficient effect on the quality of the scattering curve.

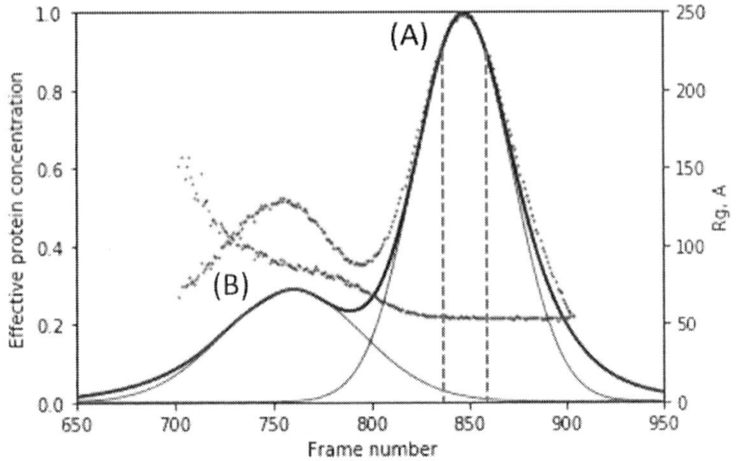

Figure 1. The curve of normalized absorbance at λ = 280 nm (blue line) corresponds to effective protein concentration, SAXS intensity at zero angle is represented in red dots. Gyration radii are represented in purple dots. Gaussian fits of two peaks are coloured in green lines. Vertical dash lines represent range of frames used for merging into final SAXS curve of the protein.

2.3. SAS data processing

Here is a brief description of data treatment, a full description can be found elsewhere, for example, in [53]. For monodisperse particles, the scattering intensity is written as:

$$I(Q) = nS(Q)\left|F(Q)\right|^2 \qquad (1)$$

where Q is the scattering vector, n is a concentration, S(q) is the structure factor and F(q) the form factor of the proteins. EID method is extrapolation I(q) vs q^2 to zero for series of scattering curves for different concentrations and gives zero concentration curve free of clustering effects. Scattering volume can be determined by [53]:

$$V = \frac{2\pi^2 I(0)}{\int Q^2 I(Q)dQ} \qquad (2)$$

For shape determination the protein is represented by a set of homogeneous atoms (beads) where the difference between the computed scattering and the experimental data is minimized [56]. Algorithm considers both scattering intensity pattern and pair-distance distribution function during operation. Best models with connectivity are averaged during processing to give the final *ab initio* model.

In this investigation integral parameters of apoferritin are calculated by ATSAS and visualised with SasView software packages [57,58]. The pair-distance distribution function is obtained with the program PRIMUS using GNOM [59]. 3D ab-initio model of apoferritin monomer is calculated by the DAMMIF program from the ATSAS-online platform.

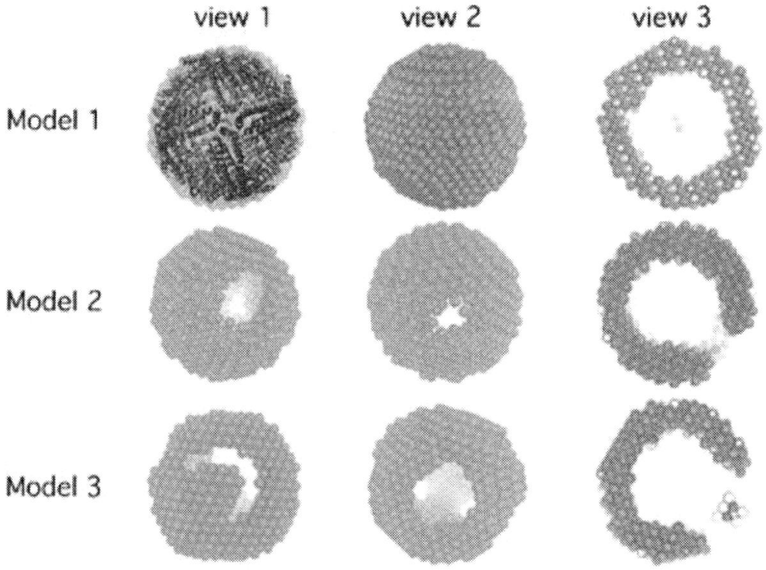

Figure 2. Projections of apoferritin dummy atom models reconstructed from SAXS and SANS data. Forward and backward projections and central section of SEC-SAXS, EID SAXS and EID SANS dummy atom models are presented as Model 1, Model 2 and Model 3 respectively. Projection of SEC-SAXS model with overlaid PDB structure (PDB code: 2W0O) is presented.

It is worth pointing out that we did a number of EID SAXS experiments and only the experiments that were done with specific pH and salt conditions gave results similar to those obtained by SEC-SAXS. For example, such a good correspondence between Rg obtained from SEC-SAXS and EID SAXS became possible only at very specified pH and salt conditions (150 mM NaCl and 5 mM Na_2HPO_4, pH = 7.3) [60], but in other experiments the discrepancies in Rg was more than 10%. It proves that if there is a sample that is not possible to be measured by SEC-SAXS (for example colloidal or magnetic particles solutions, dendrimers, powders, etc.), it strongly requires a very specific buffer conditions under which the aggregation diminishes and it became possible to apply EID SAXS.

3. Results and Discussion

Here we compare scattering data obtained by SEC-SAXS method with data obtained by the EID method from our previous SAXS and SANS experiments also. The full description of these experiments is available in [60]. During the subsequent analysis of scattering data obtained with EID method are denoted as EID SAXS and EID SANS data for X-ray and neutron data, respectively.

3.1. Core shell model analysis and pair-distribution function

Dummy atom models calculated during processing the SEC-SAXS curve are shown in Fig. 3 and core shell model fit of SAXS curve is shown in Fig. 4. In particular, the scattering curve is fitted by a core shell model from SasView program [58] at q-range 0.007-0.305 $Å^{-1}$ and structural analysis using ATSAS package was done [57]. Apoferritin PDB structure (PDB code: 2W0O) was processed by PEPSI-SAXS program and fitted to experimental data (Fig. 5) [61]. The results of chi-squared goodness of fit test are $chi^2 = 47.6$ and $chi^2 = 51.1$ for SEC-SAXS and EID SAXS models, respectively. We suggest this result is mainly due to poor fit in the Q-region near the tails of the curves. Such divergence may correspond to higher packing of crystal apoferritin in comparison to apoferritin in solution. PEPSI-SAXS fit is presented on Kratky plot (IQ^2 vs Q) (Fig. 5) and Guinier plot (ln(I) vs Q^2) (Fig. 6). The radius of gyration and scattering volume of apoferritin were directly estimated with PRIMUS. Comparing the previous results obtained by SANS and SAXS EID [60] and SEC-SAXS, it is shown that parameters such as radius of gyration, inner and outer radii are almost the same for both methods. However, the scattering volume gives different results for all models. Possible reasons of the difference obtained in SANS and SAXS experiments may be caused by the hydration shell, which is not detected by SANS, or because of hydrogen/deuterium exchange or sample impurities, for example, as reported in [62]. In any case, the difference between inner radii of apoferritin obtained by SAXS and SANS is similar to the difference between outer radii obtained by SAXS and SANS (~2.5 Å). This fact could imply that the same reasons are responsible for the difference in inner and outer apoferritin compartments. Difference in volumes obtained in SAXS EID and SEC-SAXS could occur due to the reasons mentioned above or aggregation effects or structural ordering [63]. It is known that the EID method could not exclude oligomerization and therefore scattering volume will give overestimated results.

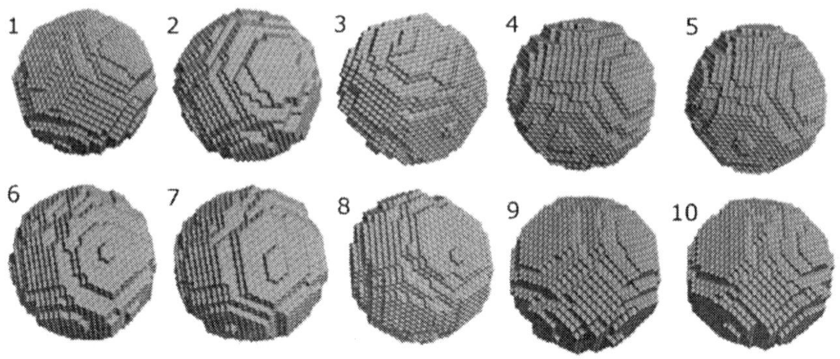

Figure 3. Dummy atom models calculated during processing SEC-SAXS curve used for averaging into final model. All intermediate models have nearly spherical form and cavity inside.

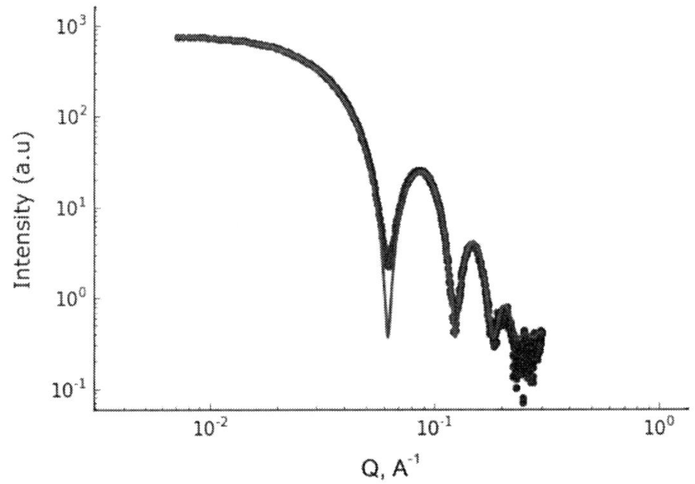

Figure 4. SEC-SAXS scattering curve is presented in blue dots, a curve of core shell model fit is presented in green line. Q-range is 0.006 – 0.2 1/Å

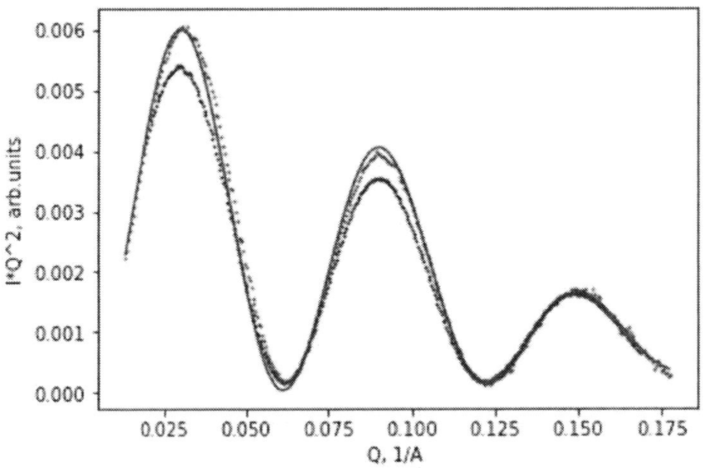

Figure 5. Kratky plot (IQ^2 versus Q) of SEC-SAXS and EID SAXS curves overlaid with PEPSI-SAXS fit. Blue dots represent EID, red dots – SEC-SAXS experimental curves. Green curve represents PEPSI-SAXS fit of experimental curves. Curves presented within the Q-range 0.01 - 0.175 1/Å.

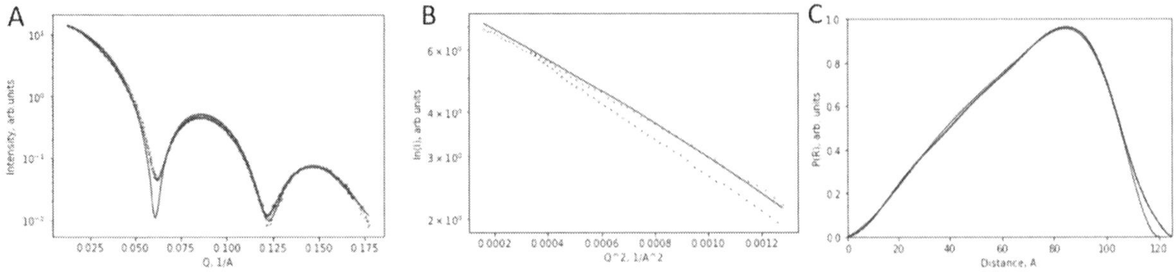

Figure 6. Intensity (A) and Guinier (B) and pair-distance plots of SEC-SAXS (red dots) and EID SAXS (blue dots) are presented and overlaid with CRYSOL fit (green line). Curves presented on Intensity plot (A) within Q-range 0.02 – 0.175 1/Å. Curves on Guinier plot (B) (ln(I) versus Q2) are presented within Q-range 0.01 – 0.035 1/Å, that includes Q-region corresponding to QRg < 1 condition. Distance distribution functions (C) of corresponding SEC-SAXS (green line) and EID SAXS (blue line) curves are presented.

The pair-distance distribution function P(r) was obtained by indirect Fourier transformation of apoferritin experimental SAXS data using GNOM program and is shown in Fig. 6. This function allows the calculation of the largest dimension in a particle. In particular, for apoferritin monomer D_{max} is 120 Å that almost corresponds to the outer diameter in terms of spherical shell model (D = 123 Å) and previously obtained results (D = 125 Å). We assume that the difference between D_{max} and D could appear because apoferritin shell is not completely spherical in solution as it is not completely rigid.

3.2. Core dummy atom model analysis

Projections of apoferritin low-resolution 3D models are shown in Fig 2. All presented models have almost spherical surfaces with cavities that corresponds to the literature data [62] and PDB structure (2W0O). Dummy atom model obtained by SEC-SAXS corresponds to PDB (Fig. 2). Models obtained by EID SAXS and SANS demonstrate statistical artefacts of data analysis process (cavities in surface). The difference to the data between the fit of the full sphere and a sphere with the hole to the data is minimal. The hole is a statistical result of the random seed. In any models with the holes during the merging process the hole will be align and persist. Despite PEPSI-SAXS fit for both SEC-SAXS and EID SAXS data results in similar curves and both curves lead to almost similar gyration radius, the 1D curves show slight difference and predict different scattering volume. We think this difference may show that SEC-SAXS is better for excluding clustering effects than EID SAXS and SEC-SAXS does not match the scattering volume (predicted by PEPSI-SAXS) because PEPSI-SAXS does not take into account thermal fluctuations of apoferritin under native conditions. We also note that the SAXS dummy atom model predicts slightly larger outer radii and scattering volume than the SANS model. It might be explained by deuterium substitution effect and difference in scattering densities obtained by SAXS and SANS [53].

Conclusion

We have determined a 3D ab initio model of apoferritin using SEC-SAXS and compared this model with PDB and models obtained with EID SAXS and EID SANS. Both EID and SEC-SAXS methods give representatives and complementary results, so whenever possible SEC-SAXS should be undertaken as well as EID (for higher concentrations) and comparisons of the data should be made to validate the data analysis process and increase confidence in the models and conclusions made from the data. We assume that data obtained by SEC-SAXS may be more suitable than EID SAXS data for making SAXS combined with X-ray diffraction or cryo-EM data analysis.

Acknowledgments
We acknowledge the European Synchrotron Radiation Facility for providing synchrotron radiation facilities. We thank ESRF structural biology group for their assistance during the experiments. The investigation used the platforms of the Grenoble Instruct Centre (ISBG; UMS 3518 CNRS-CEA-UJF-EMBL). We also thank Russian Science Foundation (grant No. 16-15-00242); FRISBI (grant No. ANR-10-INSB-05-02); GRAL within the Grenoble Partnership for Structural Biology (PSB) (grant No. ANR-10-LABX-49-01); CEA(IBS) and HGF(FZJ) (contract No. CEA(IBS) – HGF(FZJ) STC 5.1 specific agreement).

References
[1] Hura G L, Menon A L, Hammel M, Rambo R P, Poole F L, Tsutakawa S E, Jenney F E, Classen S, Frankel K A, Hopkins R C, Yang S-J, Scott J W, Dillard B D, Adams M W W and Tainer J A 2009 Robust, high-throughput solution structural analyses by small angle X-ray scattering (SAXS). *Nat. Methods* **6** 606–12
[2] Kuklin A I, Islamov A K and Gordeliy V I 2005 Scientific Reviews: Two-Detector System for Small-Angle Neutron Scattering Instrument *Neutron News* **16** 16–8
[3] Rambo R P and Tainer J A 2013 Accurate assessment of mass, models and resolution by small-angle scattering. *Nature* **496** 477–81
[4] Kuklin A I, Rogachev A V, Soloviov D V, Ivankov O I, Kovalev Y S, Utrobin P K, Kutuzov S A, Soloviev A G, Rulev M I and Gordeliy V I 2017 Neutronographic investigations of supramolecular structures on upgraded small-angle spectrometer YuMO *J. Phys. Conf. Ser.* **848** 12010
[5] Bulavin L A, Gordeliy V I, Ivan'kov O I, Islamov A K and Kuklin A I 2010 Neutron studies of the NaBr impurity influence on micelle formation in the heay water-tetradecyltrimethylammonium bromide systems *Ukr. J. Phys.* **55** 288–92
[6] Bulavin L A, Ivankov O I, Islamov A K and Kuklin A I 2010 Structure transformations in the triple liquid system tetradecyltrimethylammonium bromide-D2O-NaBr *Ukr. J. Phys.* **55** 410–4
[7] Rogachev A V, Cherny A Y, Ozerin A N, Muzafarov A M, Atatarinova E, Islamov A K, Gordeliy V I and Kuklin A I 2008 Revealing inner structure of the polycarbosilane dendrimers from small-angle neutron scattering data *J. Phys. Conf. Ser.* **129** 12041
[8] Anghel L, Balasoiu M, Ishchenko L A, Stolyar S V, Rogachev A V, Kurkin T S, Kuklin A I, Raikher Y L, Iskhakov R S and Arzumanian G M 2012 SAXS Studies of Ultrasonicated Dispersions of Biomineral Particles Produced by <i>Klebsiella oxytoca</i> *Solid State Phenom.* **190** 621–4
[9] Anitas E M, Cherny A Y, Osipov V A and Kuklin A I 2014 Small-angle scattering from three-phase systems: Investigation of the crossover between mass fractal regimes *J. Phys. Conf. Ser.* **490** 12028
[10] Gibhardt H, Haramagatti C R, Islamov A K, Ivankov O I, Kuklin A I and Eckold G 2015 Ordering Phenomena in Surfactant Systems: From Micellar Solutions to Gel and Crystalline Phases *Zeitschrift für Phys. Chemie* **229** 1869–85
[11] Soloviov D, Zabashta Y, Bulavin L, Olexandr I, Gordeliy V and Kuklin A 2014 Changes in the Area per Lipid Molecule by P-V-T and SANS Investigations *Macromol. Symp.* **335** 58–61
[12] Solov'ev D V., Kuklin A I, Utrobin P K, Ozerin A N, Kurkin T S, Ivan'kov O I, Bulavin L A and Gordelii V I 2011 X-ray scattering and volumetric P-V-T studies of the dimyristoylphosphatidylcholine-water system *J. Surf. Investig. X-ray, Synchrotron Neutron Tech.* **5** 7–10
[13] Balasoiu-Gaina A-M, Balasoiu M, Ivankov O, Soloviov D, Lysenko S, Stan C, Lupu N, Creanga D and Kuklin A 2017 Structural analysis of aqueous ferrofluids with cobalt ferrite particles stabilized with lauric acid and sodium n-dodecyl sulphate *J. Phys. Conf. Ser.* **848**

12026
[14] Shibaev A V, Makarov A V, Aleshina A L, Rogachev A V, Kuklin A I and Philippova O E 2017 Structure and oil responsiveness of viscoelastic fluids based on mixed anionic/cationic wormlike surfactant micelles *J. Phys. Conf. Ser.* **848** 12019
[15] Shibaev A V, Kuklin A I and Philippova O E 2017 Effect of polymer on the arrangement of mixed anionic/cationic wormlike surfactant micelles revealed by SANS *J. Phys. Conf. Ser.* **848** 12006
[16] Dragolici A C, Balasoiu M, Orelovich O L, Ionascu L, Nicu M, Soloviov D V, Kuklin A I, Lizunov E I and Dragolici F 2017 CEM V based special cementitious materials investigated by means of SANS method. Preliminary results *J. Phys. Conf. Ser.* **848** 12024
[17] Soloviov D V, Bulavin L A, Gordeliy V I, Gorshkova Y E, Ivankov O I, Kovalev Y S, Kuklin A I and Nikolaenko T Y Neutron Scattering Investigations of the Lipid Bilayer Structure Pressure Dependence *Nucl. Phys. At. energy* **13** 83–8
[18] Egorov V V, Zabrodskaya Y A, Lebedev D V, Gorshkov A N and Kuklin A I Structural features of the ionic self-complementary amyloidogenic peptide *J. Phys. Conf. Ser.* **848** 12022
[19] Loiko P A, Rachkovskaya G E, Skoptsov N A, Arzumanyan G M, Kulik M, Kuklin A I, Zakharevich G B, Yumashev K V. and Mateos X 2017 Mechanisms of Up-Conversion Luminescence in Glass-Ceramics Containing Er:PbF2 Nanocrystals *J. Appl. Spectrosc.* **84** 194–201
[20] Kulvelis Y V, Ivanchev S S, Primachenko O N, Lebedev V T, Marinenko E A, Ivanova I N, Kuklin A I, Ivankov O I and Soloviov D V. 2016 Structure and property optimization of perfluorinated short side chain membranes for hydrogen fuel cells using orientational stretching *RSC Adv.* **6** 108864–75
[21] Murugova T N, Gordeliy V I, Kuklin A I, Kovalev Y S, Yurkov V I, Nurenberg A, Islamov A K and Yaguzhinskii L S 2006 Detection of new double-membrane structures in native mitochondria by the method of small-angle neutron scattering *Biophysics (Oxf).* **51** 882–6
[22] Murugova T N, Gordeliy V I, Kuklin A I, Solodovnikova I M and Yaguzhinsky L S 2007 Study of three-dimensionally ordered structures of intact mitochondria by small-angle neutron scattering *Crystallogr. Reports* **52** 521–4
[23] Murugova T N, Solodovnikova I M, Yurkov V I, Gordeliy V I, Kuklin A I, Ivankov O I, Kovalev Y S, Popov V I, Teplova V V. and Yaguzhinsky L S 2011 Potentials of Small-angle Neutron Scattering for Studies of the Structure of "Live" Mitochondria *Neutron News* **22** 11–4
[24] Zabrodskaya Y A, Lebedev D V., Egorova M A, Shaldzhyan A A, Shvetsov A V., Kuklin A I, Vinogradova D S, Klopov N V., Matusevich O V., Cheremnykh T A, Dattani R and Egorov V V. 2018 The amyloidogenicity of the influenza virus PB1-derived peptide sheds light on its antiviral activity *Biophys. Chem.* **234** 16–23
[25] Selivanova O M, Surin A K, Ryzhykau Y L, Glyakina A V., Suvorina M Y, Kuklin A I, Rogachevsky V V. and Galzitskaya O V. 2018 To Be Fibrils or To Be Nanofilms? Oligomers Are Building Blocks for Fibril and Nanofilm Formation of Fragments of Aβ Peptide *Langmuir* **34** 2332–43
[26] Ryzhykau Y, Nikolaev M, Zabelskii D, Kuklin A and Gordeliy V 2017 Trimers of dimers of SRII/HtrII full complex. Small angle scattering structural investigation *FEBS J.* **284** 154
[27] Kadochnikov V V., Egorov V V., Shvetsov A V., Kuklin A I, Isaev-Ivanov V V. and Lebedev D V. 2016 Modeling of conformational transitions of fibrillogenic peptide, homologous to beta-domain of human alpha-lactalbumin *Crystallogr. Reports* **61** 98–105
[28] Egorov V V., Shaldzhyan A A, Gorshkov A N, Zabrodskaya Y A, Lebedev D V., Kuklin A I, Ksenofontova O I, Shvetsov A V., Vasin A V., Tsybalova L M and Isaev-Ivanov V V. 2016 On the structural features of influenza A nucleoprotein particles from small-angle X-ray scattering data *J. Surf. Investig. X-ray, Synchrotron Neutron Tech.* **10** 322–5
[29] Schmidt A E, Shvetsov A V., Kuklin A I, Lebedev D V., Surzhik M A, Sergeev V R and Isaev-Ivanov V V. 2016 Small-angle scattering study of Aspergillus awamori glycoprotein

glucoamylase *Crystallogr. Reports* **61** 149–52
[30] Franke D and Svergun D I 2009 DAMMIF , a program for rapid ab-initio shape determination in small-angle scattering *J. Appl. Crystallogr.* **42** 342–6
[31] Zimm B H 1948 The Scattering of Light and the Radial Distribution Function of High Polymer Solutions *J. Chem. Phys.* **16** 1093
[32] Lambright D, Malaby A W, Kathuria S V., Nobrega P R, Osman B, Matthews R C, Muthurajan U, Luger K, Chopra R, Irving T C and Chakravarthy S 2013 COMPLEMENTARY TECHNIQUES ENHANCE THE QUALITY AND SCOPE OF INFORMATION OBTAINED FROM SAXS *Trans. Am. Crystallogr. Assoc.* **44**
[33] Mathew E, Mirza A, Menhart N and IUCr 2004 Liquid-chromatography-coupled SAXS for accurate sizing of aggregating proteins *J. Synchrotron Radiat.* **11** 314–8
[34] David G, Pérez J and IUCr 2009 Combined sampler robot and high-performance liquid chromatography: a fully automated system for biological small-angle X-ray scattering experiments at the Synchrotron SOLEIL SWING beamline *J. Appl. Crystallogr.* **42** 892–900
[35] Graewert M A, Franke D, Jeffries C M, Blanchet C E, Ruskule D, Kuhle K, Flieger A, Schäfer B, Tartsch B, Meijers R and Svergun D I 2015 Automated pipeline for purification, biophysical and x-ray analysis of biomacromolecular solutions. *Sci. Rep.* **5** 10734
[36] Watanabe Y and Inoko Y 2009 Size-exclusion chromatography combined with small-angle X-ray scattering optics *J. Chromatogr. A* **1216** 7461–5
[37] Grant T D, Luft J R, Wolfley J R, Tsuruta H, Martel A, Montelione G T and Snell E H 2011 Small angle X-ray scattering as a complementary tool for high-throughput structural studies *Biopolymers* **95** 517–30
[38] Acerbo A S, Cook M J, Gillilan R E and IUCr 2015 Upgrade of MacCHESS facility for X-ray scattering of biological macromolecules in solution *J. Synchrotron Radiat.* **22** 180–6
[39] Brennich M E, Round A R and Hutin S 2017 Online Size-exclusion and Ion-exchange Chromatography on a SAXS Beamline. *J. Vis. Exp.*
[40] Russo C J and Passmore L A 2014 Ultrastable gold substrates for electron cryomicroscopy. Supplementary *Science* **346** 1377–80
[41] Heger Z, Skalickova S, Zitka O, Adam V and Kizek R 2014 Apoferritin applications in nanomedicine. *Nanomedicine (Lond).* **9** 2233–45
[42] Jurado R, Castello F, Bondia P, Casado S, Flors C, Cuesta R, Dominguez-Vera J M, Orte A and GALVEZ N 2016 Apoferritin Fibers: a New Template for 1D Fluorescent Hybrid Nanostructures *Nanoscale* 9648–56
[43] Andrews S C 1998 Iron Storage in Bacteria *Adv. Microb. Physiol.* **40** 281–351
[44] Crichton R R and Ward R J 1998 Iron homeostasis. *Met. Ions Biol. Syst.* **35** 633–65
[45] Massover W H 1993 Ultrastructure of ferritin and apoferritin: A review *Micron* **24** 389–437
[46] Ford G C, Harrison P M, Rice D W, Smith J M, Treffry A, White J L and Yariv J 1984 Ferritin: design and formation of an iron-storage molecule. *Philos. Trans. R. Soc. Lond. B. Biol. Sci.* **304** 551–65
[47] Kanekiyo M, Wei C-J, Yassine H M, McTamney P M, Boyington J C, Whittle J R R, Rao S S, Kong W-P, Wang L and Nabel G J 2013 Self-assembling influenza nanoparticle vaccines elicit broadly neutralizing H1N1 antibodies *Nature* **499** 102–6
[48] Geninatti Crich S, Cadenazzi M, Lanzardo S, Conti L, Ruiu R, Alberti D, Cavallo F, Cutrin J C and Aime S 2015 Targeting ferritin receptors for the selective delivery of imaging and therapeutic agents to breast cancer cells *Nanoscale* **7** 6527–33
[49] Sahandi Zangabad P, Karimi M, Mehdizadeh F, Malekzad H, Ghasemi A, Bahrami S, Zare H, Moghoofei M, Hekmatmanesh A and Hamblin M R 2017 Nanocaged platforms: modification, drug delivery and nanotoxicity. Opening synthetic cages to release the tiger *Nanoscale* **9** 1356–92
[50] Stevens R G, Jones D Y, Micozzi M S and Taylor P R 1988 Body Iron Stores and the Risk of Cancer *N. Engl. J. Med.* **319** 1047–52

[51] Vlasov A, Kovalev Y, Ryzhykau Y, Frolov F, Zinovev E, Rogachev A, Kuklin A and Gordeliy V 2016 Light-induced electrical properties of bacteriorhodopsin in purple membranes *FEBS J.* **283** 218
[52] Vlasov A V, Kovalev Y S, Utrobin P K, Ryzhykau Y L, Frolov F V, Zinovev E V, Rogachev A V, Kuklin A I and Gordeliy V 2017 Photo-voltage of highly-oriented bacteriorhodopsin in purple membranes: Possibilities for bio solar cells *Optoelectron. Adv. Mater. Commun.* **11** 65–7
[53] Feigin L A and Svergun D I 1987 *Structure Analysis by Small-Angle X-Ray and Neutron Scattering - Springer* ed George W. Taylor
[54] Pernot P, Round A, Barrett R, De Maria Antolinos A, Gobbo A, Gordon E, Huet J, Kieffer J, Lentini M, Mattenet M, Morawe C, Mueller-Dieckmann C, Ohlsson S, Schmid W, Surr J, Theveneau P, Zerrad L and McSweeney S 2013 Upgraded ESRF BM29 beamline for SAXS on macromolecules in solution *J. Synchrotron Radiat.* **20** 660–4
[55] Brennich M E, Kieffer J, Bonamis G, De Maria Antolinos A, Hutin S, Pernot P and Round A 2016 Online data analysis at the ESRF bioSAXS beamline, BM29 *J. Appl. Crystallogr.* **49** 203–12
[56] Volkov V and Svergun D I 2003 Uniqueness of ab-initio shape determination in small-angle scattering *J. Appl. Crystallogr.* **36** 860–4
[57] Petoukhov M V., Franke D, Shkumatov A V., Tria G, Kikhney A G, Gajda M, Gorba C, Mertens H D T, Konarev P V. and Svergun D I 2012 New developments in the ATSAS program package for small-angle scattering data analysis *J. Appl. Crystallogr.* **45** 342–50
[58] Butler P, Alina G, Hernandez R, Doucet M and A. J 2013 SASView for Small Angle Scattering Analysis
[59] Svergun D I 1992 Determination of the regularization parameter in indirect-transform methods using perceptual criteria *J. Appl. Crystallogr.* **25** 495–503
[60] Murugova T N, Vlasov A V, Ivankov O I, Rogachev A V, Soloviov D V, Zhigunov A, Kovalev Y S, Ryzhykau Y L, Zinovev E V, Round A, Gordeliy V I and Kuklin A I 2015 Low resolution structural studies of apoferritin via SANS and SAXS: The Effect of concentration *J. Optoelectron. Adv. Mater.* **17** 1397–402
[61] Grudinin S, Garkavenko M and Kazennov A 2017 Pepsi-SAXS : an adaptive method for rapid and accurate computation of small-angle X-ray scattering profiles *Acta Crystallogr. Sect. D Struct. Biol.* **73** 449–64
[62] de Val N, Declercq J-P, Lim C K and Crichton R R 2012 Structural analysis of haemin demetallation by L-chain apoferritins *J. Inorg. Biochem.* **112** 77–84
[63] Vlasov A, Murugova T, Grudinin S, Ivankov O, Soloviov D, Rogachev A, Round A, Ryzhykau Y, Mishin A, Balandin T, Borschevskiy V, Gordeliy V and Kuklin A 2014 Protein structure and structural ordering versus concentration dependence *FEBS J.* **281** 593

Institute of Physics
Dirac House, Temple Back
Bristol BS1 6BE UK

ISSN: 1742-6588
ISBN 978-1-5108-6071-1

CURRAN ASSOCIATES INC.
proceedings.com

VI School on Cosmic Rays and Astrophysics 2015

Journal of Physics: Conference Series Volume 866

Chiapas, Mexico
17 - 25 November 2015

IOP
Institute of Physics